Anonymous

The Arctic

A history of its discovery, its plants, animals and natural phenomena

Anonymous

The Arctic

A history of its discovery, its plants, animals and natural phenomena

ISBN/EAN: 9783337228798

Printed in Europe, USA, Canada, Australia, Japan

Cover: Foto ©berggeist007 / pixelio.de

More available books at **www.hansebooks.com**

A HISTORY OF ITS DISCOVERY,

ITS PLANTS, ANIMALS, AND NATURAL PHENOMENA,

SPITZBERGEN ISLAND.

EDINBURGH
GEORGE TOD, ANNANDALE STREET
1876

PREFACE.

ENGLISHMEN have always felt a special interest in the regions of the icy North, from the days when Dr. Thorne first proposed the search after a passage to the Pole, down to these present times, when the solution of the problem seems to be near at hand. The interest originally kindled by commercial considerations has been maintained by purer and loftier motives,—by the thirst after knowledge, and the sympathy with the brave deeds of brave men. And it must be admitted that our national virtues of resolute perseverance and patient courage have never been more happily displayed than in the prosecution of the great work of Arctic Discovery. Our explorers have refused to know when they were beaten; and in defiance of a terrible climate, of icebergs and ice-floes, of hurricanes and driving snow-storms, of obstacles, dangers, and difficulties, have pressed onward, until the latest adventurers stand on the threshold of the unknown region,—on the very shore of the great sea that, in all probability, surrounds the Pole. Their labours, indeed, have been attended by the shadows of melancholy disasters, and the long Arctic night closes over the graves of many whom England was loath to lose; but in their successful issue they have brought us acquainted with the phenomena of a strange and wonderful world, and opened up to us a succession of scenes of the most remarkable character.

There can be no question that in the frozen wastes and snowy wildernesses lurks a powerful fascination, which proves almost irresistible to the adventurous spirit. He who has once entered the Arctic World, however great his sufferings, is restless until he returns to it. Whether the spell lies in the weird magnificence of the scenery, in the splendours of the heavens, in the mystery which still hovers over those far-off seas of ice and remote bays, or in the excitement of a continual struggle with the forces of Nature, or whether all these influences are at work, we cannot stop to inquire. But it seems to us certain that the Arctic World has a romance and an attraction about it, which are far more powerful over the minds of men than the rich glowing lands of the Tropics, or the

"Summer-isles of Eden lying in dark-purple spheres of sea,"

which are crowned with the bread-fruit and the palm, the spontaneous gifts of a liberal soil. We follow with far deeper interest the footprints of a Parry and a Franklin than those of a Wallis, a Carteret, or even a Cook.

The general reader, therefore, may not be displeased at the attempt of the present writer to put before him, with bold touches, and in outline rather than in detail, a picture of that Polar World which is so awful and yet so fascinating. In the following pages he will find its principal features sketched, its chief characters legibly and clearly traced. They are not intended for the scientific,—though it is hoped the scientific, if they fall in with them, will find no ground for censure. They aim at describing the wonders of sky and sea and land; the glories of the aurora; the beauty of the starry Arctic night; the majesty of iceberg and glacier; the rugged dreariness of the hummocky fields of ice; the habits of the Polar bear, the seal, and the walrus; and the manners and customs of the various tribes which frequent the shores of the Polar seas and straits, or dwell on the border-land of the Frigid Zone. In a word, it has been the writer's object to bring together just such particulars as might enable the intelligent reader to realize to himself the true character of the world which extends around the North Pole. In carrying out this object, he has necessarily had recourse to the voyages of numerous explorers and the narratives of sundry scientific authorities; and he believes that not a statement has been ventured which could not claim their support.

CONTENTS.

CHAPTER I.

Various routes between the Atlantic and Pacific Oceans described—Advantages of a North-West Passage, if practicable—What is to be gained from further Arctic exploration—What zoology would gain—The problem of the migration of birds—About the Knots—Boundaries of the North Polar Regions—Their principal geographical features—Divisions into two zones, or sections—The stony tundras—The flora of the North—The Siberian desert—Limits of perpetual snow—General character of life in the Polar World..............9-21

CHAPTER II.

An imaginary voyage—View of the Greenland coast—A splendid picture of land and sea—The winter night and its atmospheric phenomena—The aurora borealis described—Its peculiarities and possible causes—Winds and whirlwinds—Phenomena of refraction—The "ice-blink"—Characteristics of the Arctic night—Described by Dr. Kane—Remarkable atmospheric conditions—Effect of prolonged darkness on animal life—Characteristics of the Arctic spring—A spring landscape described by Dr. Hayes—Summer in the North—The Northern heavens and the Pole-Star—List of Northern constellations—The Great Bear—Some conspicuous stars......22-40

CHAPTER III.

The Polar seas—Formation of icebergs—Their dimensions and appearance—Description of colossal bergs—Their danger to navigation—Adventures with bergs—Quotations from various writers—Dissolution of an iceberg—Icebergs in Melville Bay—How icebergs are formed—Reference to icebergs in the Alpine lakes—Professor Tyndall quoted—Breaking up of a berg described by Dr. Hayes—A vision of icebergs—Their range—The "pack-ice" described—Extent of the ice-fields—"Taking the pack"—An incident described by Admiral Beechey—Dangerous position of Captain Parry's ships—Character of an ice-field—Crossing an ice-field—Its extraordinary dimensions—Animal life in the Polar seas—Walrus-hunting—Quotation from Mr. Lamont—A disagreeable process—Natural history of the walrus—The walrus and the Polar bear—Historical sketch of the walrus-fishery—Adventure with walruses—A walrus-hunt described—Hunting in an Arctic gale—The Phocidæ family—Natural history of the seal—Different genera—Seal's flesh, and its uses—An incident in Dr. Kane's expedition—An Eskimo hut—An Eskimo seal-hunter—The whale, and all about it—The Greenland whale—What is whalebone?—Food of the whale—The Northern rorqual—Eskimo whale-fishers—About the narwhal—The black dolphin—The orc, or grampus—The Polar bear—Bears and seals—Particulars of the habits of the Polar bear—His voracity—Affection of the bear for her young—An episode described—Battle with a bear—The bear and the Eskimo dogs—The Arctic night—Its various phases—Coming of the sun—Return of the birds—Guillemots and auks—About the puffins—The mergansers—The smew, or white nun—The eider duck described—Eider ducks in Iceland—Collecting eider down—The wild swan—Fables about its death-song—The Arctic waters, and their teeming life—Migrations of fish..41-107

CHAPTER IV.

The formation of snow described—Snow-crystals—Effects of the crystallizing force—Ice-flowers—Sir David Brewster's experiment with polarised light—Regelation and moulding of ice—Characteristics of glacier-ice—Cleavage in compact ice—The aspect of glaciers—On the motion of glaciers—History of its discovery—Moraines described—Theory of glacier-motion—Quotation from Professor Tyndall—Glaciers of the Polar Regions—Glacier in Bell Sound—Formation of icebergs—Icebergs in Baffin Bay—Glacier described by Dr. Hayes—The Greenland Mer de Glace—Glacier of Sermitsialik—The great Humboldt Glacier—Discovered by Dr. Kane—Description of its features—Kane's theory of icebergs—Notes on the glacier..108-134

CHAPTER V.

Red snow, what is it?—First forms of vegetable life—The lichens, their variety—Reindeer moss—Rock-hair—Rock tripe, or *tripe de roche*—Used as food—Iceland moss and its properties—The mosses of the Arctic Regions—Scurvy-grass—The fly-agaric—Microscopic vegetation—A memorial of Franklin—Phænogamous plants of the North—Cryptogamous plants—Vegetation in Novaia Zemlaia—In Spitzbergen—In Kamtschatka—The *Fritollaria sarrana*—The wooded and desert zones—Forms of animal life—Natural history of the reindeer—His usefulness—His food—Reindeer and wolves—Cunning of the Arctic wolf—Domesticity of the wolf—The musk-ox described—Captain M'Clintock quoted—The Arctic fox—His wariness—A fox-trap—The bear and the fox—The Arctic hare—The Alpine hare—The Hudson Bay lemming—The Mustelidæ family—The marten—The sable—The polecat—About the glutton, or wolverine—anecdotes of his extraordinary sagacity—A great enemy to the trapper—The biter bit—Arctic birds—The falcons—The crows—Distribution of animals..............................135-161

CHAPTER VI.

Iceland, its extent—Its history—Its volcanoes—Hekla and its eruptions—Eruption of the Skaptá Jokul—The geysers, or boiling springs—Their phenomena described—Account of the Strokr—Coasts and valleys of Iceland—The Thingvalla—Description of Reikiavik, the capital—Character of the Icelander—His haymaking operations—His dwelling described—An Icelandic Church—Icelandic clergy—Travelling in Iceland—Its inconveniences—Fording the streams—Fishing in Iceland..........................162-174

CHAPTER VII.

The land of the Eskimos—Range of the so-called Arctic Highlanders—Danish settlements in Greenland—Upernavik described—Jacobshav'n—Godhav'n—Their Eskimo inhabitants—The Moravian missions—Characteristics of the nomadic Eskimos—Their physical qualities—Their mode of dress—An Eskimo hut—The Eskimo *kayak*, or canoe—Their weapons and implements—Hostility between the Eskimos and Red Indians—Eskimo settlement at Anatoak—Eskimo singing—Food of the Eskimos—Dr. Hayes' intercourse with the Eskimos—The story of Hans the Hunter—The Eskimo dogs—Anecdote of Toodla—The Eskimo sledge—Equipment of the sledge—Equipment of an Eskimo hunter—General character of the Eskimos..175-196

CHAPTER VIII.

Lapland, its divisions, extent, and boundaries—Its climate—Its inhabitants—Their physical characteristics—Dress of the Lapps—Their superstitions—The Mountain Lapps—Their migratory habits—Their *tuguria*, or huts, described—Milking the reindeer—Sledging and skating—A Lapp's skates—A Lapp's sledge—The Lapp hunters—Encounter with a bear—Intemperance of the Mountain Lapps—The Forest Lapps—Interior economy of a Lapland hut—Lapps at Bjorkholm—Racial characteristics of the Lapps—Habits and manners of the Lapps—The Lapp dialect—The Lapps and the Quénes—The stationary Lapps, and their *gârds*.....197-207

CHAPTER IX.

The Samojedes—Their degrading superstitions—Samojede idol at Waigatz—The *Tadebtsios*, or spirits—Influence of the *Tadibe*, or sorcerer—His mode of incantation—Customs of the Samojedes—The Ostiaks—Their *Schaitans* and *Schamans*—Residence of the Ostiaks—Hunting the white bear—Kamtschatka described—Its inhabitants—Their physical peculiarities—The dog of Kamtschatka—His qualities—His usefulness—How he is trained—Siberia and its tribes—The Jakuts—Their *jerts*, or huts—Their hardy horses—The character of the Jakuts—Jakut travellers—Jakut merchants and their caravans—Dreariness of the country they inhabit—Hunting the reindeer—At Kolymsk—The Tungusi—His mode of travelling—His food—The Tchuktche, and their land—Their activity as traders—Tobacco, a staple of commerce—Visit to a Tchuktche family—The Tenngyk and the Oukilon............208-221

CHAPTER X.

History of Discovery in the Arctic Regions—Expeditions of Thorne and Hore—Of Sir Hugh Willoughby—Martin Frobisher and his adventures—Death of Sir Humphrey Gilbert—Discoveries of Davis—Hudson, his discovery of Hudson Bay, Jan Mayen, and Cape Wolstenholm—His fate—Baffin's voyages—Highway to the North Pole—Expedition of Ross and Parry—Parry's second expedition—Loss of the *Fury*—Overland journeys—Franklin's last expedition—The search after Franklin—Discovery of relics—Captain Penny's expedition—The graves on Beechey Island—Sir Robert M'Clure's discovery of the North-West Passage—Dr. Rae obtains intelligence of the fate of Franklin—Voyage of M'Clintock—Further memorials—Lieutenant Hobson's discoveries—Dr. Kane's expedition—Explores Smith Sound—Discovers the Humboldt Glacier and Kennedy Channel—His misfortunes—Wintering in the Arctic Regions—Dr. Hayes' expedition—Voyage of the *Germania* and the *Hansa*—Loss of the latter—Escape of the crew on an ice-raft—A coal-hut—Arrival at Greenland—Adventures of the *Germania*—Barents and Carlsen—Austrian expedition under Payer—Voyage of the *Polaris*, and her fate—English expedition of 1875.....222-273

List of Illustrations.

1. THE CREW OF THE "HANSA" DRAGGING THEIR BOATS ACROSS THE ICE (FRONTISPIECE).
2. A DESERT OF ICE IN THE ARCTIC REGION, 13
3. THE SWAMPS OF THE OBI, ... 16
4. IN THE FOREST ZONE OF THE NORTH (FULL-PAGE), ... 17
5. THE MIDNIGHT SUN (FULL-PAGE), ... 23
6. OFF THE COAST OF GREENLAND, ... 25
7. MOONLIGHT IN THE POLAR WORLD, ... 26
8. THE AURORA BOREALIS, ... 28
9. THE AURORA BOREALIS—THE CORONA, ... 29
10. ATMOSPHERIC PHENOMENA IN THE ARCTIC REGIONS:—REFLECTION OF ICEBERGS, ... 32
11. ADVENT OF SPRING IN THE POLAR REGIONS, 35
12. URSA MAJOR AND URSA MINOR, ... 36
13. NEBULA IN ANDROMEDA, ... 39
14. ARCHED ICEBERG OFF THE GREENLAND COAST, 42
15. AMONG THE BERGS—A NARROW ESCAPE, ... 43
16. ICEBERG AND ICEFIELD, MELVILLE BAY, GREENLAND, ... 45
17. ORIGIN OF ICEBERGS—EXTENSION OF A GLACIER SEAWARDS, 47
18. THE ALETSCH GLACIER, SWITZERLAND, FROM THE ÆGGISCHHORN, SHOWING ITS MORAINES, ... 48
19. THE MARJELEN SEA, SWITZERLAND, ... 49
20. FALL OF AN ICEBERG (FULL-PAGE), 51
21. IN AN ICE-PACK, MELVILLE BAY, ... 53
22. CHANNEL IN AN ICE-FIELD, 54
23. "NIPPED" IN AN ICE-FIELD, ... 54
24. AMONG THE ICE-HUMMOCKS (FULL-PAGE), ... 57
25. HUNTING THE WALRUS, ... 61
26. THE WALRUS, OR MORSE, ... 63
27. A WALRUS FAMILY, ... 64
28. FIGHT BETWEEN A WALRUS AND A POLAR BEAR, ... 64
29. BOAT ATTACKED BY A WALRUS (FULL-PAGE), ... 65
30. FIGHT WITH A WALRUS, ... 68
31. HERD OF SEALS, NEAR THE DEVIL'S THUMB, BAFFIN SEA, GREENLAND, ... 71
32. THE COMMON SEAL, ... 73
33. SHOOTING A SEAL, ... 74
34. THE OTARY, 75
35. THE HOODED SEAL, ... 76
36. AN ESKIMO SEAL-HUNTER, 77
37. THE GREENLAND WHALE, 79
38. NARWHALS, MALE AND FEMALE, ... 82
39. A SHOAL OF DOLPHINS, ... 83
40. POLAR BEARS, ... 84
41. BEAR CATCHING A SEAL, ... 86
42. BEARS DESTROYING A CACHE, ... 88
43. FIGHT WITH A WHITE BEAR (FULL-PAGE), 89
44. STALKING A BEAR, ... 94
45. SEA-BIRDS IN THE POLAR REGIONS, 97
46. THE AUK, ... 98
47. PUFFINS, ... 99
48. THE GOOSANDER, ... 100
49. A BIRD "BAZAAR" IN NOVAIA ZEMLAIA (FULL-PAGE), 101
50. THE BLACK-BACKED GULL, 103
51. THE EIDER-DUCK, ... 103
52. THE HAUNT OF THE WILD SWAN, ... 105
53. VARIOUS FORMS OF SNOW-CRYSTALS, ... 109
54. EXHIBITION OF ICE-FLOWERS BY PROJECTION, ... 110
55. ICE-FLOWERS, ... 110
56. MOULDING ICE, ... 112
57. A POLAR GLACIER, ... 118
58. GLACIER, ENGLISH BAY, SPITZBERGEN, ... 119
59. GLACIER, BELL SOUND, SPITZBERGEN, ... 120
60. STEAMER "CHARGING" AN ICEBERG, UPERNAVIK, GREENLAND (FULL-PAGE), ... 121
61. FORCING A PASSAGE THROUGH THE ICE (FULL-PAGE), ... 125
62. THE GLACIER OF SERMIATSIALIK, GREENLAND (FULL-PAGE), ... 129
63. PROTOCOCCUS NIVALIS, ... 130
64. WILD REINDEER, ... 145
65. THE MUSK-OX, ... 150
66. ARCTIC FOXES, ... 152
67. A FOX-TRAP, ... 153
68. THE ERMINE, OR SABLE MARTEN, ... 156
69. THE GLUTTON, OR WOLVERINE, 157
70. PTARMIGAN, ... 160
71. AN ICELANDIC LANDSCAPE, 163

72. MOUNT HEKLA, FROM THE VALLEY OF HEVITA, 164
73. THE GREAT GEYSER, 166
74. HARBOUR OF REIKIAVIK, ... 169
75. ICELANDERS FISHING FOR NARWHAL, 174
76. UPERNAVIK, GREENLAND, 176
77. DISCO ISLAND, GREENLAND, ... 177
78. GODHAV'N, DISCO ISLAND, GREENLAND, ... 177
79. DANISH SETTLEMENT OF JACOBSHAV'N, GREENLAND, 178
80. BUILDING AN ESKIMO HUT, ... 181
81. THE ESKIMO KAYAK, 182
82. THE ESKIMO OOMIAK, ... 183
83. DR. HAYES FALLS IN WITH HANS THE HUNTER (FULL-PAGE), 187
84. ESKIMO DOGS, ... 191
85. ESKIMO SLEDGE AND TEAM (FULL-PAGE), ... 193
86. REINDEER IN LAPLAND, ... 200
87. TRAVELLING IN LAPLAND, ... 201
88. FISHER LAPPS, ... 203
89. SAMOJEDE HUTS ON WAIGATZ ISLAND, ... 209
90. A SAMOJEDE FAMILY, ... 210
91. JAKUT HUNTER AND BEAR, ... 212
92. KAMTSCHATKANS, ... 213
93. A KAMTSCHATKAN SLEDGE AND TEAM, ... 215
94. THE LOSS OF THE "SQUIRREL," 224
95. SHIP OF THE SEVENTEENTH CENTURY, ... 225
96. SCENERY OF JAN MAYEN, ... 226
97. THE "HECLA" AND "FURY" WINTERING AT WINTER ISLAND, 229
98. THE "FURY" ABANDONED BY PARRY, ... 230
99. DISCOVERY OF THE CAIRN CONTAINING SIR JOHN FRANKLIN'S PAPERS, ... 233
100. RELICS OF THE FRANKLIN EXPEDITION BROUGHT BACK TO ENGLAND, ... 235
101. DISCOVERY OF ONE OF THE BOATS OF THE FRANKLIN EXPEDITION, ... 236
102. THE "THREE BROTHER TURRETS," 238
103. MORTON ON THE SHORE OF THE SUPPOSED POLAR OCEAN, ... 240
104. DR. KANE PAYING A VISIT TO AN ESKIMO HUT AT ETAH, ... 241
105. THE CREW OF THE "HANSA" TRYING TO LASSO A BEAR, (FULL-PAGE) 247
106. THE MIDNIGHT SUN, GREENLAND, 249
107. A BEAR AT ANCHOR, ... 249
108. SKATING—OFF THE COAST OF GREENLAND, ... 250
109. SNOW LINNETS AND BUNTINGS VISITING THE CREW OF THE "HANSA," 254
110. THE CREW OF THE "HANSA" BIVOUACKING ON THE ICE, (FULL-PAGE) 255
111. A RASH INTRUDER, ... 259
112. BEAR-HUNTING—GREENLAND, 260
113. "INTO A WATER-GAP," 261
114. THE CREW OF THE "GERMANIA" IN A SNOW-STORM (FULL-PAGE), ... 263
115. MATERIALS FOR THE HOUSE, ... 266
116. ATTACK ON A BEAR. 267
117. SETTING FOX-TRAPS, ... 2[illegible]
118. RELIEVED, ... 297

THE ARCTIC WORLD.

CHAPTER I.

THE NORTH POLE—THRESHOLD OF THE UNKNOWN WORLD—THE CIRCUMPOLAR REGIONS—THE FLORA OF THE NORTH—LIFE IN THE POLAR WORLD—THE NORTH-WEST AND NORTH-EAST PASSAGES.

S the reader knows, the Poles are the two extremities of the axis round which the Earth revolves. It is to the North Pole, and the regions surrounding it, that the following pages will be devoted.

The inhabitants of Western Europe, and more particularly those of the British Isles, have a peculiar interest in the North Polar Regions. Deriving their wealth and importance from their commercial enterprise, and that commercial enterprise leading their ships and seamen into the furthest seas, they have necessarily a vital concern in the discovery of the shortest possible route from that side of the Earth which they inhabit to the other, or eastern side; and this, more particularly, because the East is rich in natural productions which are of high value to the peoples of the West.

Now a glance at the map will show the reader that the traders of Western Europe—the British, the French, the Dutch, the Scandinavians—are situated on the northern shores of the Atlantic Ocean, and that, to reach the Pacific Ocean or the Indian, only two routes are at present open. For instance, they may cross the Atlantic to the American coast, and, keeping southward, strike through Magellan's stormy Strait or round the bleak promontory of Cape Horn into the Pacific, and then, over some thousands of miles of water, proceed to Australia or Hindustan or China; or they may keep along the African coast to the Cape of Good Hope, its southernmost point, and so stretch across the warm Tropical seas to India and the Eastern Archipelago. A third, an *artificial* route, has indeed of late years been opened up; and ships, entering the Mediterranean, may pass through the Suez Canal into the Red Sea. But this last-named route is unsuitable for sailing-ships, and all three routes are laborious and slow. How greatly the distance would be shortened were it possible to navigate the Northern Seas, and, keeping along the north coast of the American continent, to descend Behring's Strait into the Pacific! In other words, were that North-West Passage practicable, which, for three centuries, our geographers and explorers so assiduously and courageously toiled to discover! But a still shorter route would be opened up, if we could follow a line drawn from the British Islands straight across the North Pole to Behring's Sea and the Aleutian Archipelago. This line would not exceed 5000 miles in length, and would bring Japan, China, and India within a very short voyage from Great Britain. We should be able to reach Japan in three or four weeks, to the obvious advantage of our extensive commerce.

Hitherto, however, all efforts to follow out this route, and to throw open this great ocean-highway between Europe and Asia, have failed. Man has been baffled by Nature; by ice, and frost, and winds, and climatic influences. With heroic perseverance he has sought to gain the open sea which, it is believed, surrounds the Pole, but a barrier of ice has invariably arrested his progress. His researches have carried him within about 500 miles of the coveted point; but he is as yet unable to move a step beyond this furthest limit of geographical discovery. Immediately around the North Pole, within a radius of eight to ten degrees or more, according to locality, still lies an Unknown Region, on the threshold of which Science stands expectant, eagerly looking forward to the day when human skill and human courage shall penetrate its solitudes and reveal its secrets.

This Unknown Region comprises an area of 2,500,000 square miles; an immense portion of the terrestrial surface to be shut out from the knowledge of Civilized Man. Its further exploration, if practicable, cannot but be rich in valuable results. Not only would it furnish the shortest route from the West to the East, from progressive Europe to conservative Asia, from the Atlantic to the Pacific, but it could not fail to add in a very important degree to our stores of scientific information. Sir Edward Sabine is surely right when he says, that it is the greatest geographical achievement which can be attempted, and that it will be the crowning enterprise of those Arctic researches in which England has hitherto had the pre-eminence.

We may briefly indicate to the reader some of the advantages which might be expected from exploration in the Unknown Region. It would unquestionably advance the science of hydrography, and lead to a solution of some of the more difficult problems connected with the Equatorial and Polar ocean-currents, those great movements of the waters of which, as yet, we know so little.

A series of pendulum observations, it is said, at and near the North Pole, would be of essential service to the science of geology. We are unable, at present, for want of sufficient data, to form a mathematical theory of the physical condition of the Earth, and to ascertain its exact configuration. No pendulum observations have been taken nearer than 600 or 620 miles to the North Pole.

Again: what precious information respecting the strange and wonderful phenomena of magnetism and atmospheric electricity would certainly be acquired! How much we have yet to learn in reference to the Aurora, which can be learned only in high latitudes, and at or near the point which apparently represents a magnetic focus or centre!

It has also been pointed out by Mr. Markham that the climate of Europe is largely affected by the atmospheric conditions of the Polar area, in which the development of extremely low temperatures necessarily leads to corresponding extreme changes of pressure, and other atmospheric disturbances, whose influence extends far into the Temperate Zone. For the satisfactory appreciation of these phenomena, says Mr. Markham, a precise knowledge is required of the distribution of land and water within the Polar Region; and any addition to our knowledge of its unknown area, accompanied by suitable observations of its meteorology, cannot fail to afford improved means of understanding the meteorology of our own country, and of the Earth generally.

There can be no doubt, too, that geology would profit, if we could push our researches nearer to the Pole, and force our way through the great barrier of the Polar ice. It is highly desirable, too, that we should know more of that interesting class of animals, the Mollusca, both terrestrial and aquatic, fresh-water and salt-water. Again: what a wide field of inquiry is opened up by the Polar glaciers; their extent, their elevation, their range, and the effects produced by the slow but continuous motion of those huge ice-rivers over the surface of the country. And the botanist has a right to calculate upon the discovery of many precious forms of vegetable life in the Unknown Region. The Arctic flora is by no means abundant, but it is peculiarly interesting. In Greenland, besides numerous mosses, lichens, algæ, and the like, flourish three hundred kinds of flowering plants, all of which are natives of the Scandinavian peninsula; and Dr. Joseph Hooker remarks that they exhibit scarcely any admixture of American types, though these are found on the opposite coast of Labrador. It would seem probable that in the warm period which preceded the Glacial Age, the Scandinavian flora spread over the entire area of the Polar Regions; but that during the Age of Ice it was gradually driven within its present limits, only the hardier types surviving the blight of the long lingering winter.

And what would be the gain to the zoologist? Why, it is a well-known fact that life abounds in the Arctic waters, and especially those minute organisms which play so important a part in the formation of sedimentary deposits, and help to build up the terrestrial crust. We have much to learn, moreover, of the habits and habitats of the fish, the echinoderms, the molluscs, the corals, the sponges of the extreme Northern Seas.

There are questions connected with the migrations of birds which can be elucidated only by an exploration of the Unknown Region. Multitudes which annually visit our shores in the winter and spring, return in summer to the far North. This is their regular custom, and obviously would not have become a custom unless it had been found beneficial. Therefore we may assume that in the zone they frequent they find some water which is not always frozen; some land on which they can rest their weary feet; and an adequate supply of nourishing food.

From Professor Newton we adopt, in connection with this consideration, a brief account of the movements of one class of migratory birds,—the Knots.*

The knot, or sandpiper, is something half-way between a snipe and a plover. It is a very active and graceful bird, with rather long legs, moderately long wings, and a very short tail. It swims admirably, but is not often seen in the water; preferring to assemble with its fellows on the sandy sea-shores, where it gropes in the sand for food, or fishes in the rock-pools and shallow waters for the small crustaceans. It is known both as the red and the ash-coloured sandpiper, because it changes the colour of its plumage according to the season of the year; a bright red in summer, a sober ashen-gray in winter. Now, in the spring the knot seeks our island in immense flocks, and after remaining on the coasts for about a fortnight, can be traced proceeding gradually northwards, until it finally takes leave of us. It has been noticed in Iceland and Greenland, but not to stay; the summer there would be too rigorous for its liking, and it goes further and

* The *Tringa canutus* of ornithologists.

further north. Whither? Where does it build its nest, and hatch its young? We lose all trace of it for some weeks: what becomes of it?

Towards the end of summer back it comes to us in larger flocks than before, and both old birds and young birds remain upon our coasts until November, or, in mild seasons, even later Then it wings its flight to the south, and luxuriates in blue skies and balmy airs until the following spring, when it resumes the order of its migrations.

Commenting upon these facts, Professor Newton infers that the lands visited by the knot in the middle of summer are less sterile than Iceland or Greenland; for certainly it would not pass over these countries, which are known to be the breeding-places for swarms of water-birds, to resort to regions not so well provided with supplies of food. The *food*, however, chiefly depends on the *climate*. Wherefore we conclude that beyond the northern tracts already explored lies a region enjoying in summer a climate more genial than they possess.

Do any races of men with which we are now unacquainted inhabit the Unknown Region? Mr. Markham observes that although scarcely one-half of the Arctic world has been explored, yet numerous traces of former inhabitants have been found in wastes which are at present abandoned to the silence and solitude. Man would seem to migrate as well as the inferior animals, and it is possible that tribes may be dwelling in the mysterious inner zone between the Pole and the known Polar countries.

The extreme points reached by our explorers on the ice-bound Greenland coast are in about 82° on the west, and 76° on the east side; these two points lying about six hundred miles apart. As man has dwelt at both these points, and as they are separated from the settlements further south by a dreary, desolate, uninhabitable interval, it is not an extravagant conjecture that the unknown land to the north has been or is inhabited. In 1818 a small tribe was discovered on the bleak Greenland coast between 76° and 79° N.; their southward range being bounded by the glaciers of Melville Bay, and their northward by the colossal mass of the Humboldt Glacier, while inland their way is barred by the *Sernik-sook*, a great glacier of the interior. These so-called Arctic Highlanders number about one hundred and forty souls, and their existence "depends on open pools and lanes of water throughout the winter, which attract animal life." Wherever such conditions as these are found, man may be found.

We know that there are or have been inhabitants north of the Humboldt Glacier, on the very threshold of the Unknown Region; for Dr. Kane's expedition discovered the runner of a sledge made of bone lying on the beach immediately to the north of it. The Arctic Highlanders, moreover, cherish a tradition that herds of musk-oxen frequent an island situated far away to the north in an iceless sea. Traces of these animals were found by Captain Hall's expedition, in 1871–72, as far north as 81° 30′; and similar indications have been noted on the eastern side of Greenland. In 1823, Captain Clavering found twelve natives at Cape Borlase Warren, in lat. 79° N.; but when Captain Koldewey, of the German expedition, wintered in the same neighbourhood, in 1869, they had disappeared, though there were traces of their occupancy, and ample means of subsistence. Yet they cannot have gone southward, owing to insuperable natural obstacles; they must have moved towards the North Pole.

We have thus indicated some of the results which may be anticipated from further researches in the Unknown Region. It is not to be forgotten, however, that "the unexpected always happens," and it is impossible to calculate definitely the consequences which may ensue from a more extensive investigation. "Columbus," it has been justly said, "found very few to sympathize with him, or perceive the utility of the effort on his part to go out into the unknown waste of waters beyond the Strait of Gibraltar, in search of a new country. Who can, at this time, estimate the advantages which have followed upon that adventure? If now it should be possible to reach the Pole, and to make accurate observations at that point, from the relation which the Earth bears to the sun and to the whole stellar universe, the most useful results are very likely to follow, in a more thorough knowledge of our globe."

The reader has now before him the particulars which will enable him to form an idea of the extent and character of the undiscovered region of the Pole. Roughly speaking, it is bounded by the 80th parallel of latitude on the European side, except at a few points where our

A DESERT OF ICE IN THE ARCTIC REGION.

gallant explorers have succeeded in crossing the threshold; on the Asiatic side it descends as low as 75°; and to the west of Behring Strait as low as 72°. Thus, it varies from 500 or 600 to 1400 or 1500 miles across. Below these parallels, and bounded by the Arctic Circle, or, in some places, by the 60th parallel, extends a vast belt of land and water which is generally known as the Arctic or Circumpolar Regions. These have been more or less thoroughly explored; and it is to a description of their principal features, their forms of animal and vegetable life, and their natural phenomena, that we propose to devote the present volume.

It is important to remember that the northern shores of Europe, Asia, and America are skirted by the parallel of 70°, and that the belt between the 70th and 80th parallels, having been partially explored by the seamen and travellers of various nations, intervenes as a kind of neutral ground between the known and the unknown. We may, indeed, formulate our statement thus;

from the Pole to the 80th degree stretches the unknown; from the 80th to the 70th, the partially known; while, south of the 70th, we traverse the lands and seas which human enterprise has completely conquered.

The Circumpolar Zone includes the northernmost portions of the three great continents, Europe, Asia, and America; and by sea it has three approaches or gateways: one, through the Northern Ocean, between Norway and Greenland; another, through Davis Strait,—both these being from the Atlantic; and a third, through Behring Strait,—the entrance from the Pacific.

It will be seen that the Circumpolar Regions, as they are now understood, and as we shall describe them in the following pages, extend to the south of that imaginary line drawn by geographers round the North Pole, at a distance from it equal to the obliquity of the ecliptic, or 23° 30′. Within this circle, however, there is a period of the year when the sun does not set; while there is another when he is never seen, when a settled gloom spreads over the face of nature,—this period being longer or shorter at any given point according as that point is nearer to or further from the Pole.

But as animal and vegetable life are largely affected by climate, it may be justly said that wherever an Arctic climate prevails *there* we shall find an Arctic or Polar region; and, hence, many countries below even the 60th parallel, such as Kamtschatka, Labrador, and South Greenland, fall within the Circumpolar boundary.

The waters surrounding the North Pole bear the general designation of the Arctic Ocean. But here again it is almost impossible to particularize any uniform limit southward. It joins the Pacific at Behring Strait in about lat. 66° N., and consequently in this quarter extends fully half a degree *beyond* the Arctic Circle. At Scoresby Sound, as at North Cape, where it meets the Atlantic, it is intersected by the parallel of 71°, and consequently falls short of the Arctic Circle by about 4° 30′.

In the Old World, the Polar Ocean, if we include its gulfs, extends, in the White Sea, fully two degrees *beyond* the Arctic Circle; while at Cape Severo, the northernmost point of Asia, in lat. 78° 25′ N., it is 11° 55′ distance from it. Finally, in the New World it is everywhere confined *within* the Circle; as much as 5° at Point Barrow, about 7° 30′ at Barrow Strait, and about 3° at the Hecla and Fury Strait.

We may add that, so far as temperature is concerned, the great gulfs known, in memory of their discoverers, as Davis Strait, Baffin Bay, and Hudson Bay, are portions of the Arctic Ocean.

Of the more southerly area of this great ocean, the only section which has been adequately explored to a distance from the continent, and in the direction of the Pole, is that which washes the north-east of America. Here we meet, under the collective name of the Polar Archipelago, with the following islands:—Banks Land, Wollaston Land, Prince Albert Land, Victoria Land, Prince Patrick Island, Princess Royal Islands, Melville Island, Cornwallis Island, North Devon, Beechey Island, Grinnell Land, and North Lincoln. Further to the east lie Spitzbergen, Jan Mayen Island, Novaia Zemlaia, New Siberia, and the Liakhov Islands. The chief straits and inlets are Lancaster Sound, Barrow Strait, Smith Sound, Regent Inlet, Hecla and Fury Strait, Wellington Channel, and Cumberland Sound; while further westward are Belcher Channel, Melville Sound, M'Clintock Channel, Banks Strait, and Prince of Wales Strait.

The Arctic Lands comprehend two well-defined sections, or zones; that of the forests, and the treeless wastes.

To the latter belong the islands within the Arctic Circle, and also a considerable tract of the northern continents, forming the "barrens" of North America, and the "tundras" and "steppes" of European Russia and Siberia.

The treeless character of this vast area of wilderness is owing to the bleak sea-winds which drive, without let or hindrance, across the islands and level shores of the Polar Ocean, compelling even the most vigorous plant to bend before them and creep along the ground.

Drearier scenes are nowhere presented than these stony tundras, or their boundless swamps. Almost the only vegetation are a few gray lichens, a few dull blackish-looking mosses; the stunted flowers or crawling grasses that here and there occur do not relieve the uniform desolation,—they serve simply to enhance its gloomy character. In summer, indeed, the tundras are full of life; for the spawning instinct of the salmon and the sturgeon impels them to enter their rivers and seek the quiet recesses of their mysterious lakes. The reindeer assemble in numerous herds to feed on the herbage warmed into temporary vitality by the upward-slanting sun; the whirr of countless wings announces the coming of the migratory birds to breed, and feed their young, on the river-banks and the level shores; and in their trail arrive the eagle and the hawk, intent on prey.

But with the first days of September a change passes over the scene. Animal life hastens to the more genial south; the birds abandon the frozen wastes; the reindeer retires to the shelter of the forests; the fish desert the ice-bound streams; and a terrible silence reigns in the desolate wilderness, broken only by the harsh yelp of a fox or the melancholy hooting of a snow-owl. For some eight or nine months a deep shroud or pall of snow lies on the whitened plains. No cheerful sunbeams irradiate it with a rosy glow; the sky is dull and dark; and it seems as if Nature had been abandoned to eternal Night.

But blank and dreary as the limitless expanse of snow appears, it is the security of man in these far northern regions. It affords the necessary protection to the scanty vegetable life against the rigour of the long winter season. In Rensselaer Bay, Dr. Kane found, when the surface temperature had sunk to $-30°$, a temperature at two feet deep of $-8°$, at four feet deep of $+2°$, and at eight feet deep of $+26°$, or no more than 6° below freezing-point. Hence, underneath their thick frozen pall, the Arctic grasses and lichens maintain a struggling existence, and are able to maintain it until thoroughly resuscitated by the summer sun. It is owing to this wise and beneficent provision that, in the highest latitudes, the explorer discovers some feeble forms of vegetation. Thus, as Hartwig reminds us, Morton gathered a crucifer at Cape Constitution, in lat. 80° 45′ N.; and Dr. Kane, on the banks of the Minturn River, in lat. 78° 52′, met with a flower-growth which, though fully Arctic in its type, was gaily and richly coloured—including the purple lychnis, the starry chickweed, and the hesperis, among the festuca and other tufted grasses.

In the tundras, the most abundant vegetable forms, next to the lichens and mosses, are the grasses, the crucifers, the saxifrages, the caryophyles, and the compositæ. These grow fewer and fewer as we move towards the north, but the number of individual plants does not decrease. Where the soil is fairly dry, we shall find an extensive growth of lichens; in moister grounds, these are intermingled with the well-known Iceland moss. Lichens are everywhere, except in the sparse tracts of meadow-land lying at the foot of sheltering hills, or in those alluvial inundated hollows which are thickly planted with "whispering reeds" and dwarf willows.

It is not easy to trace exactly the boundary between the tundras and the forest zone. The former descend to the south, and the latter advances to the north, according to the climatic influences which prevail; following the isothermic lines of uniform temperature, and not the mathematical limits of the geographical parallels of latitude. Where the ground undulates, and hilly ridges break the fury of the icy blasts, the forests encroach on the stony treeless region; but the desolate plains strike into the wooded zone in places where the ocean-winds range with unchecked sway.

The southernmost limit of the "barrens" is found in Labrador, where they descend to lat. 57°; nor is this to be wondered at, when we remember the peculiar position of that gloomy peninsula, with icy seas washing it on three sides, and cold winds sweeping over it from the north. On the opposite coasts of Hudson Bay they do not strike lower than 60°; and they continue to

THE SWAMPS OF THE OBI.

rise as we proceed westward, until in the Mackenzie Valley we find the tall forest growth reaching as far north as 68° or even 70°. Thence they recede gradually, until, on the bleak shore of Behring Sea, they do not rise higher than 65°. Crossing into the eastern continent, we find them beginning, in the land of the Tuski (or Tchuktche), in 63°, and from thence encroaching gradually upon the tundras until, at the Lena, they reach as high as 71°. From the Lena to the Obi the tundras gain upon the forests, and in the Obi Valley descend below the Arctic Circle: but from the Obi to the Scandinavian coast the forests gain upon the tundras, terminating, after many variations, in lat. 70°.

The result to which this rapid survey brings us is, that the "tundras" or "barrens" of Europe, Asia, and America occupy an area larger than the whole of Europe. The Siberian wilderness is more extensive than the African Sahara or the South American Pampas. But of still vaster area are the Arctic forest regions, which stretch in an "almost continuous belt"

IN THE FOREST ZONE OF THE NORTH.

through three quarters of the world, with a breadth of from 15° to 20°—that is, of 1000 to 1400 miles. And it is a peculiarity of these Circumpolar woods, that they are almost wholly composed of conifers, and that frequently a wide space of ground is covered for leagues upon leagues with a single kind of fir or pine.

> "This is the forest primeval. The murmuring pines and the hemlocks,
> Blended with moss, and in garments green, indistinct in the twilight,
> Stand like Druids of eld, with voices sad and prophetic,
> Stand like harpers hoar, with beards that rest on their bosoms."

The American species, however, differ from the Asiatic or European. While in the Hudson Bay territories grow the white and black spruce,* the Canadian larch,† and the gray pine;‡ in Scandinavia and Siberia, the Siberian fir and larch,§ the *Picea olovata*, and the *Pinus umbra* flourish. But both in the Old World and the New the birch advances beyond the fir and pine, and on the banks of the rivers and the shores of the lakes dwarf willows form immense and almost impenetrable thickets. The Arctic forests also include various kinds of ash, elder, and the service tree; and though orchard trees are wholly wanting, both man and beast find a great boon in the bilberries, cranberries, bog-berries, and the like, which grow plentifully in many localities.

The area of the Arctic flora comprises Greenland, in the western hemisphere, and extends considerably to the south of the Arctic Circle, especially on the coasts, where it reaches the parallel of 60° N. lat., and even overpasses it.

In Greenland the vegetation is more truly of an Arctic character than even in Iceland. The valleys are covered with marsh-plants and dingy mosses; the gloomy rocks are encrusted with lichens; while the grasses on the meadow-lands that border the fiords and inlets are nearly four times less varied than those of Iceland.

The flora of Iceland is approximative to that of Great Britain; yet only one in every four of British plants is included in it. The total number of species may be computed at eight hundred and seventy, of which more than half blossom; this proportion is greater than prevails in Scotland, but then only thirty-two are of woody texture. They are scattered about in groups according as they prefer a marshy, volcanic, dry, or marine soil. Many bloom in the immediate vicinity of the hot springs; some not far from the brink of the basin of the Great Geyser, where every other plant is petrified; and several species of confervæ flourish in a spring the waters of which are hot enough, it is said, to boil an egg.

From the nature of the Arctic forests, the reader will be prepared to learn that they are not inhabited, like those of the Tropics, by swarms of animals; or made musical by the songs of birds, like our European woods. Even the echoes are silent, except when the hoarse wind bears to them the peculiar cry of the reindeer, the howl of the wolf, or the sharp scream of some bird of prey. Insect life, however, is active and abundant; and our Arctic travellers have suffered greatly from the legions of gnats which haunt their swampy recesses.

Passing from the forest region into the treeless wastes, we may glance once again at their strikingly impressive features. North of the 62nd parallel no corn can ripen, on account of the fatal power of the winds which pour down from the Arctic Ocean. As we advance to the north-

* *Abies alba et nigra.* † *Larix Canadensis.* ‡ *Pinus Banksiana.* § *Abies Sibirica, Larix Sibirica.*

ward, a wide-spread area of desolation stretches before us: salt steppes, stony plains, boundless swamps, and lakes of salt and fresh water. So terrible is the cold that the spongy soil is perpetually frozen to the depth of some hundred feet below the surface; and the surface itself, though not thawed until the end of June, is again ice-bound by the middle of September. One of the most graphic sketches with which we are acquainted of the extreme Siberian desert is furnished by Admiral von Wrangel, who travelled during the winter from the mouth of the Kolyma to Behring Strait.

Here, he says, endless snows and ice-crusted rocks bound the horizon; Nature lies shrouded in all but perpetual winter; life is a constant conflict with privation and with the terrors of cold and hunger; the grave of Nature, containing only the bones of another world. The people, and even the snow, throw off a continual vapour; and this evaporation is instantly changed into millions of needles of ice, which make a noise in the air like the sound of torn satin or the rustle of thick silk. The reindeer take to the forest, or crowd together for heat; and the raven alone, the dark bird of winter, still smites the frosty air with heavy laborious wing, leaving behind him a long trail of thin vapour to mark the course of his solitary flight. The trunks of the thickest trees are rent with a loud clang, masses of rock are torn from their sites, the ground in the valleys is split into a myriad fissures, from which the waters that are underneath bubble up, throwing off a cloud of smoke, and immediately congealing into ice. The atmosphere grows dense; the glistening stars are dimmed. The dogs outside the huts of the Siberians burrow in the snow, and their howling, at intervals of six or eight hours, interrupts the general silence of winter.

The abundance of fur-bearing animals in the less rigorous parts of the tundras has induced the hardy Russians to colonize and build towns on these confines of the Frozen World. Yakutsk, on the river Lena, in 62° 1′ 30″ N., may be regarded, perhaps, as the coldest town on the Earth. The ground is perpetually frozen to the depth of more than 400 feet, of which three feet only are thawed in summer, when Fahrenheit's thermometer frequently marks 77° in the shade. Yet in winter the rigour of the climate is so extreme that mercury is constantly frozen for two and occasionally even for three months.

From the data set forth in the preceding pages, the reader will conclude that, as indeed results from physical laws, the line of perpetual snow will be found to descend lower and lower on advancing to the Pole. By the line of perpetual snow we mean, of course, the limit above which a continual frost endures. Now, this limit varies according to climate. The lower the temperature, the lower the snow-line; the higher the temperature, the higher the snow-line. In the Tropics it does not sink below the summits of the loftiest mountains. Thus, at 1° from the Equator, where the mean temperature at the sea-level is 84°.2, the snow-line must be sought at the elevation of 15,203 feet; in 51° 30′ lat., the latitude of London, it is usually found at about 5900 feet; in lat. 80°, where the mean temperature is 33°.6, it sinks to 457 feet. These figures, however, represent its *normal* elevations; but temperature, as we all know, is greatly affected by local circumstances, and therefore the perpetual snow-line varies greatly in height. Owing to causes already explained, the snow-line in the Circumpolar Regions sinks to a very low level; and, therefore, many mountainous regions or elevated table-lands, such as Spitzbergen, Greenland, and Novaia Zemlaia, which, in a more temperate climate, would bloom with emerald slopes and waving woods, are covered with huge glaciers and fields of ice, with apparently interminable reaches of untrodden snow.

It should be noted, however, that nowhere does the perpetual snow-line descend to the water's edge; nowhere has the spell of winter absolutely crushed the life out of all vegetation. Lichens and grasses, on which the reindeer gains its hardy subsistence, are found near lat. 80°; even on the awful plains of Melville Island the snow melts at midsummer; and the deserts of New Siberia afford food for considerable numbers of lemmings. As far as man has reached to the north, says a popular and accurate writer, vegetation, when fostered by a sheltered situation and the refraction of solar heat from the rocks, has everywhere been found to rise to a considerable altitude above the level of the sea; and should there be land at the North Pole, we may reasonably suppose that it is destitute neither of animal nor vegetable life. It would be quite wrong to conclude that the cold of winter invariably increases as we approach the Pole, the temperature of a land being controlled by many other causes besides its latitude. Even in the most northern regions visited by man, the influence of the sea, particularly where favoured by warm currents, considerably mitigates the severity of the winter, while at the same time it diminishes the heat of summer. On the other hand, the large continental tracts of Asia or America that slope towards the Pole, possess a more rigorous winter and a fiercer summer than many coast lands or islands situated far nearer to the Pole. For example: the western shores of Novaia Zemlaia, fronting a wide expanse of sea, have an average winter temperature of only – 4°, and a mean summer temperature which rises very little above the freezing-point of water (+ 36° 30′); while Yakutsk, situated in the centre of Siberia, and 20° nearer to the Equator, has a winter temperature of – 36° 6′, and a summer of + 66° 6′.

But though such are the physical conditions of the Polar Regions, it must not be supposed that Nature wears only a severe and repellent aspect. There is something beautiful in the vast expanse of snowy plain when seen by the light of a cloudless moon; something majestic in the colossal glaciers which fill up the remote Arctic valleys; something picturesque in the numerous icebergs which grandly sail down the dark Polar waters; something mysterious and wonderful in the coruscations of the Aurora, which illuminates the darkness of the winter nights with the glory of the celestial fires. The law of compensation prevails in the far North, as in the glowing and exuberant regions of the Tropics.

CHAPTER II.

THE ARCTIC HEAVENS: ATMOSPHERIC AND METEORIC PHENOMENA.

LET the reader fancy himself—should he be reading these pages on a warm summer's day, the fancy will not be unpleasant!—let the reader fancy himself on board a well-found, stoutly-built whaling-vessel, and rapidly approaching the coast of Greenland. But the heavy mist hangs over the legend-haunted shores, and we can but catch the sound of the clanging surf as it rolls upon them. All around us spreads the mist,—dense, impenetrable. What is that before us? The dead white mass of an iceberg, slowly drifting with the current, and almost upon us before the look-out man discovered it. But the helm has been sharply handled; our good ship has put about; and we sail clear of the mighty pyramid. Fully one hundred and fifty feet high, we can assure you, and twice as broad at its base. A sudden break in the mist reveals its radiant spire, with white cloud-wreaths circling and dancing round it in the sunlight.

And now, as we steadily move forward, the fog is lifted up like a curtain, and before us, like a scene in a panorama, looms the Greenland coast in all its austere magnificence: yonder are its broad ice-filled valleys, its snow-clad ravines, its noble mountains, its iron-bound range of cliffs, its general aspect of solemn desolation.

Away over the westward sea fly the scattered vapours, disclosing iceberg after iceberg, like the magical towers in some of Turner's pictures. We seem to have been drawn by some irresistible spell into a world of enchantment, and all the old Norse romance comes back upon the memory, with its picturesque associations. Yonder lies the Valhalla of the ancient ocean-rovers; yonder the dazzling city of the sun-god Freya, one of the most popular of the Scandinavian divinities, as well he might be; yonder the elfin caves of Alfheim; and Glitner, with its walls of gold and roofs of silver; and the radiant Gimele, the home of the blessed; and there, too, towering above the clouds, the bridge Bifrost, by which the heroes ascended from earth to heaven. Heimdall, who can see for fully a hundred leagues, as well by night as by day, stands sentinel upon it, prepared to sound his horn Gjallar, if intruders should attempt to cross it!

The sea is smooth as glass; not a ripple breaks the wonderful calmness of its surface. It is midnight, but in this strange Arctic world the sun still hangs close upon the northern horizon; the icebergs rear their dazzling crests around, like floating spires, and turrets, and many-towered minsters; the dark headlands are boldly outlined against the sky; and sea, and sky, and mountains, and icebergs are suffused in a wildly beautiful atmosphere of crimson, gold, and purple. The picture is like a poet's vision; and so startlingly unreal, that it is difficult for the unaccustomed spectator to believe it other than an illusion.

THE MIDNIGHT SUN.

We adopt the following description from the vivid language of Dr. Hayes, who displays a keen feeling for the beauties of the Polar world.

The air was warm, he says, almost as a summer's night at home, and yet there were the icebergs and the bleak mountains, with which the fancy, in our own land of green hills and waving woods, can associate nothing but what is cold and repellent. Bright was the sky, and soft and strangely inspiring as the skies of Italy. The bergs had wholly lost their chilly aspect, and, glittering in the blaze of the brilliant heavens, seemed, in the distance, like masses of burnished metal or solid flame. Nearer at hand they were huge blocks of Parian marble, encrusted with colossal gems of pearl and opal. One in particular exhibited the perfection of grandeur. Its

OFF THE COAST OF GREENLAND.

form was not unlike that of the Coliseum, and it lay so far away that half its height was buried beneath the line of blood-red waters. The sun, slow moving along its path of glory, passed behind it, and the old Roman ruin seemed suddenly to break into flame!

Nothing, indeed, but the pencil of the artist could depict the wonderful richness of this combined landscape and seascape. Church, in his great picture of "The Icebergs," has grandly exhibited a scene not unlike that we have attempted to describe.

In the shadows of the bergs the water was a rich green, and nothing could be more soft and tender than the gradations of colour made by the sea shoaling on the sloping tongues of some of these floating masses. The tint increased in intensity where the ice overhung the waters, and a deep cavern in one of them exhibited the solid colour of the malachite mingled with the transparency of the emerald, while, in strange contrast, a broad belt of cobalt blue shot diagonally through its body.

The enchantment of the scene was heightened by a thousand little cascades which flashed into the sea from the icebergs, the water being discharged from basins of melted snow and ice which tranquilly reposed far up in the hollows of their topmost surface. From other bergs large boulders were occasionally detached, and these plunged into the water with a deafening din, while the roll and rush of the ocean resounded like the music of a solemn dirge through their broken archways.

The contrasts and combinations of colour in the Polar world are, indeed, among its particular attractions, and of their kind they cannot be surpassed or imitated even in the gorgeous realms of the Tropics. The pale azure gleam of the ice, the dazzling whiteness of the snow, the vivid verdure of the sunlit plains, the deep emerald tints, crossed with sapphire and ultramarine, of the waters, would in themselves afford a multiplicity of rich and beautiful effects; but to these we must add the magical influences of the coruscations of the Arctic heavens, with the glories of the midnight sun and the wonders of the Aurora.

MOONLIGHT IN THE POLAR WORLD.

Even moonlight in the Polar world is unlike moonlight anywhere else; it has a character all its own,—strange, weird, supernatural. Night after night the sky will be free from cloud or shadow, and the radiant stars shine out with a singular intensity, seeming to cut the air like keen swords. The moonbeams are thrown back with a pale lustre by ice-floe and glacier and snow-drift, and the only relief to the brightness is where the dark cliffs throw a shadow over the landscape. Gloriously beautiful look the snow-clad mountains, as the moonlight pours upon them its serene splendour, interrupted only by the occasional passage of a wreath of mist, which is soon transformed into sparkling silver. The whole scene produces an impression of awe on the mind of the thoughtful spectator, and he feels as if brought face to face with the visible presence of another world.

The prolonged winter night is in itself well calculated to affect the imagination of the European. He reads of it in travels and books of astronomy; but to know what it *is*, and what it *means*, he must submit himself to its influence,—he must "winter" in the Polar Regions. Not to see sunrise and sunset, and the changes they bring with them, day after day, enlivening, inspiriting, strengthening, is felt at first as an intolerable burden. The stars shining at all hours with equal brilliancy, and the lasting darkness which reigns for twenty days of each winter month when the moon is below the horizon, become a weariness and a discomfort. The traveller longs for the reappearance of the moon; and yet before she has run her ten days' course, he feels fatigued by the uniform illumination.

But sometimes a relief is supplied by the phenomenon of the Aurora Borealis. We inhabitants of the United Kingdom know something of the rare beauty of the "northern lights," when the heavens kindle with a mysterious play of colours which reminds us of the strange weird radiance that occasionally kindles in our dreams; yet these are poor and trivial when compared with the auroral display. Let us endeavour to realize it from the glowing description painted by one of the most eloquent and observant of Arctic explorers.

He was groping his way among the ice-hummocks, in the deep obscurity of the mid-winter, when suddenly a bright ray darted up from behind the black cloud which lay low down on the horizon before him. It lasted but an instant, and, having filled the air with a strange illumination, it died away, leaving the darkness even greater darkness than before. Presently an arc of coloured light sprang across the sky, and the aurora became gradually more fixed. The space enclosed by the arc was very dark, and was filled with the cloud. The play of the rays which rose from its gradually brightening border was for some time very capricious, modifying the burst of flame from what seemed a conflagration of the heavens to the soft glow of early morn.

Gradually the light grew more and more intense, and from irregular bursts it settled into an almost steady sheet of splendour. This sheet, however, was far from uniform, and may best be described as "a flood of mingling and variously-tinted streaks."

The exhibition, at first tame and quiet, developed by degrees into startling brilliancy. The broad dome of night seemed all a-blaze. Lurid fires, fiercer than those which reddened the heavens from burning Troy, flashed angrily across the zenith. The stars waned before the marvellous outburst, and seemed to recede further and further from the Earth; "as when the chariot of the sun, driven by Phaeton, and carried from its beaten track by the ungovernable steeds, rushed madly through the skies, parching the world and withering the constellations. The gentle Andromeda flies trembling from the flame; Perseus, with his flashing sword and Gorgon shield, retreats in fear; the Pole-Star is chased from the night; and the Great Bear, faithful sentinel of the North, quits his guardian watch, following the feeble trail."

The colour of the light was chiefly red, but this was not permanent, and every hue mingled in the wonderful display.

Blue and yellow streamers shot athwart the lurid fire; and, sometimes starting side by side from the wide expanse of the illumined arc, they melted into each other, and flung a weird glare of green over the landscape.

Again this green overcame the red; blue and yellow blended with each other in their swift flight; violet-tinted arrows flashed through a broad glow of orange, and countless tongues of white flame, formed of these uniting streams, rushed aloft and clasped the skies. The effect of the many-coloured lustre upon the surrounding objects was singularly wonderful. The weird forms of innumerable icebergs, singly and in clusters, loomed above the sea, and around their summits hovered the strange gleam, like the fires of Vesuvius over the villas and temples of

THE AURORA BOREALIS.

Pompeii. All along the white surface of the frozen sea, upon the mountain-peaks and the lofty cliffs, the light glowed and dimmed and glowed again, as if the air were filled with graveyard meteors, flitting wildly above some vast illimitable city of the dead. The scene was noiseless, yet the senses were deceived, for sounds not of earth or sea seemed to follow the swift coruscations, and to fall upon the ear like

"The tread
Of phantoms dread,
With banner, and spear, and flame."

Though the details, so to speak, are not always the same, the general character of the aurora changes very slightly, and, from a comparison of numerous accounts, the gradation of the phenomenon would seem to be as follows:

The sky slowly assumes a tint of brown, on which, as on a background, is soon developed a nebulous segment, bordered by a spacious arc of dazzling whiteness, which seems incessantly agitated by a tremulous motion. From this arc an incredible number of shafts and rays of light leap upwards to the zenith. These luminous columns pass through all the hues of the rainbow, —from softest violet and intensest sapphire to green and purple-red. Sometimes the rays issue from the resplendent arc mingled with darker flashes; sometimes they rise simultaneously at different points of the horizon, and unite in one broad sea of flame pervaded by rapid undulations. On other occasions it would seem as if invisible hands were unfurling fiery dazzling banners, to

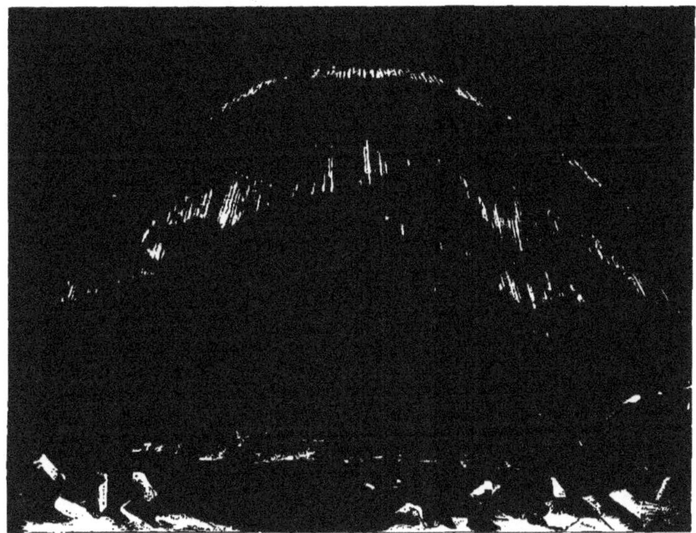

THE AURORA BOREALIS—THE CORONA.

stream, like meteors, in the troubled air. A kind of canopy, of soft and tranquil light, which is known as the *corona*, indicates the close of the marvellous exhibition; and shortly after its appearance the luminous rays begin to decrease in splendour, the richly-coloured arcs dissolve and die out, and soon of all the gorgeous spectacle nothing remains but a whitish cloudy haze in those parts of the firmament which, but a few minutes before, blazed with the mysterious fires of the aurora borealis.

The arc of the aurora is only part of a broad circle of light, which is elevated considerably above the surface of our globe, and the centre of which is situated in the vicinity of the Pole. It is not difficult, therefore, to account for the different aspects under which it is presented to observers placed at different angles to the focus of the display. A person some degrees *south* of the ring necessarily sees but a very small arc of it towards the north, owing to the interposition of the earth between him and it; if he stood nearer the north, the arc would appear larger and higher; if immediately below it, he would see it apparently traversing the zenith; or if within the ring, and still further north, he would see it culminating in the south. It has been supposed that the centre of the ring corresponds with the magnetic north pole in the island of Boothia Felix.

Generally the phenomenon lasts for several hours, and at times it will be varied by peculiar features. Now it will seem to present the hemispherical segment of a colossal wheel; now it will wave and droop like a rich tapestry of many-coloured light, in a thousand prismatic folds; and now it exhibits the array of innumerable dazzling streamers, waving in the dark and intense sky.

The arc varies in elevation, but is seldom more than ninety miles above the terrestrial surface. Its diameter, however, must be enormous, for it has been known to extend southward to Italy, and has been simultaneously visible in Sardinia, Connecticut, and at New Orleans.

According to some authorities, the phenomenon is accompanied by noises resembling the discharge of fireworks, or the crackling of silk when one piece is folded over another; but this statement is discredited by the most trustworthy observers.

Mrs. Somerville's description is worth quoting, as taking up more emphatically some points to which we have already alluded:—

The aurora, she says, is decidedly an electrical (or, more strictly speaking, a magneto-electrical) phenomenon. It generally appears soon after sunset in the form of a luminous arc stretching more or less from east to west, the most elevated point being always in the magnetic meridian of the place of the observer; across the arc the coruscations are rapid, vivid, and of various colours, darting like lightning to the zenith, and at the same time flitting laterally with incessant velocity. The brightness of the rays varies in an instant; they sometimes surpass the splendour of stars of the first magnitude, and often exhibit colours of admirable transparency,—blood-red at the base, emerald-green in the middle, and clear yellow towards their extremity. Sometimes one, and sometimes a quick succession of luminous currents run from one end of the arc or bow to the other, so that the rays rapidly increase in brightness; but it is impossible to say whether the coruscations themselves are actually affected by a horizontal motion of translation, or whether the more vivid light is conveyed from ray to ray. The rays occasionally dart far past the zenith, vanish, suddenly reappear, and, being joined by others from the arc, form a magnificent corona or immense dome of light. The segment of the sky below the arc is quite black, as if formed by dense clouds; yet M. Struve is said to have seen stars in it, and so it would appear that the blackness of which several observers speak must be the effect of contrast. The lower edge of the arc is evenly defined; its upper margin is fringed by the coruscations, their convergence towards the north, and that of the arc itself, being probably an effect of perspective.

The aurora exercises a remarkable influence on the magnetic needle, even in places where the display is not visible. Its vibrations seem to be slower or quicker according as the auroral light is quiescent or in motion, and the variations of the compass during the day show that the aurora is not peculiar to night. It has been ascertained by careful observations that the disturbances of the magnetic needle and the auroral displays were simultaneous at Toronto, in

Canada, on thirteen days out of twenty-four, the remaining days having been clouded; and contemporaneous observations show that in these thirteen days there were also magnetic disturbances at Prague and Tasmania; so that the occurrence of auroral phenomena at Toronto on these occasions may be viewed as a local manifestation connected with magnetic effects, which, whatever may have been their origin, probably prevailed *on the same day over the whole surface of the globe.*

Among the *atmospheric* phenomena of the outer world we are justified in reckoning the Winds, which are remarkable for their variability. Their force is considerably diminished when they pass over a wide surface of ice; sometimes the ice seems even to beat back the breeze, and turn it in a contrary direction. The warm airs from the south grow cool as they sweep across the frozen expanse, and give up their moisture in the form of snow. In a region so bleak and chill it is not often that clouds are created, the atmospheric vapours being condensed into snow or hail without passing through any intermediate condition

Whirlwinds of frozen snow are formidable enemies to the seaman forced to traverse the ice on foot, or in a sledge drawn by Eskimo dogs. Dense showers lash and sting the unfortunate traveller's face, penetrate his mouth and nostrils, freeze together his very eyelids, and almost blind him. His skin assumes a bluish tint, and burns as if scarred by the keen thongs of a knout.

An optical illusion of frequent occurrence in the Polar Regions makes objects appear of dimensions much larger than they really possess. A fox assumes the proportions of a bear; low banks of ice are elevated into lofty mountains. The eye is fatigued by dwelling upon the horizon of lands which are never approached. Just as in the sandy deserts of the Sahara the distances of real objects are apparently diminished, so the Arctic explorer, misled by the aërial illusion, advances towards a goal which seems always near at hand, but is never attained.

Another source of error, common both to the Arctic and the Tropical deserts, is the mirage, a phenomenon of refraction, which represents as suspended in air the images of remote objects, and thus gives rise to the most curious illusions and fantastic scenes. Dr. Scoresby one day perceived in the air the reversed representation of a ship which he recognized as the *Fame*, commanded by his father. He afterwards discovered that it had been lying moored in a creek about ten leagues from the point where the mirage had played with his imagination.

Again, in approaching a field of ice or snow, the traveller invariably descries a belt of resplendent white immediately above the horizon. This is known as the "ice-blink," and it reveals to the Arctic navigator beforehand the character of the ice he is approaching. At times, too, a range of icebergs, or of broken masses of ice, will be reflected in colossal shadows on the sky, with a strange and even weird effect.

But, after all, the special distinction between the Arctic lands and the other regions of the globe is their long day and longer night. Describing an immense spiral curve upon the horizon, the sun gradually mounts to 30°, the highest point of its course; then, in the same manner, it returns towards the horizon, and bids farewell to the wildernesses of the North, slowly passing away behind the veil of a gloomy and ghastly twilight.

When the navigator, says Captain Parry, finds himself for the first time buried in the silent shadows of the Arctic night, he cannot conquer an involuntary emotion of dread; he feels

ATMOSPHERIC PHENOMENON IN THE ARCTIC REGIONS:—REFLECTION OF ICEBERGS.

transported out of the sphere of ordinary, commonplace existence. The deadly and sombre deserts of the Pole seem like those uncreated voids which Milton has placed between the realms of life and death. The very animals are affected by the profound melancholy which saddens the face of Nature.

Who can read without emotion the following passages from Dr. Kane's Journal?—

"*October 28, Friday.*—The moon has reached her greatest northern declination of about 25° 35′. She is a glorious object; sweeping around the heavens, at the lowest part of her curve she is still 14° above the horizon. For eight days she has been making her circuit with nearly unvarying brightness. It is one of those sparkling nights that bring back the memory of sleigh-bells and songs and glad communings of hearts in lands that are far away.

"The weather outside is at 25° below zero."

A few days later, and the heroic explorer writes:—

"*November 7, Monday.*—The darkness is coming on with insidious steadiness, and its advances can be perceived only by comparing one day with its fellow of some time back. We still read the thermometer at noonday without a light, and the black masses of the hills are plain for about five hours with their glaring patches of snow; but all the rest is darkness. Lanterns are always on the spar-deck, and the lard-lamps never extinguished below. The stars of the sixth magnitude shine out at noonday.

"Our darkness has ninety days to run before we shall get back again even to the contested twilight of to-day. Altogether, our winter will have been sunless for one hundred and forty days."

Here is another significant passage; yet all its significance can scarcely be appreciated by the dwellers in temperate climes:—

"*November 27, Sunday.*—The thermometer was in the neighbourhood of 40° *below zero,* and the day was too dark to read at noon."

"*December 15, Thursday.*—We have lost the last vestige of our mid-day twilight. We cannot see print, and hardly paper: the fingers cannot be counted a foot from the eyes. Noonday and midnight are alike; and, except a vague glimmer on the sky that seems to define the hill outlines to the south, we have nothing to tell us that this Arctic world of ours has a sun."

On the 11th of January (1854), Dr. Kane's thermometer stood at 49° below zero; and on the 20th the range of those at the observatory was at – 64° to – 67°. On the 5th of February they began to show an unexampled temperature. They ranged from 60° to 75° below zero, and one very admirable instrument on the taffrail of the brig stood at – 65°. The reduced mean of the best spirit-standards gave – 67°, or 97° below the freezing-point of water.

At these temperatures chloric ether became solid, and carefully prepared chloroform exhibited a granular film or pellicle on its surface. Spirit of naphtha froze at – 54°, and oil of sassafras at – 49°. The oil of winter-green assumes a flocculent appearance at – 56°, and solid at – 63° and – 65°.

Some further details, borrowed from Dr. Kane's experiences, will illustrate still more powerfully the singular atmospheric conditions of the Arctic winter.

The exhalations from the surface of the body invested any exposed or partially-clad part with a wreath of vapour. The air had a perceptible pungency when inspired, but Dr. Kane did not undergo the painful sensation described by some Siberian travellers. When breathed for any length of time it imparted a sensation of dryness to the air-passages; and Dr. Kane observed that all his party, as it were involuntarily, breathed gradually, and with compressed lips.

It was at noon on the 21st of January that the first glimmer of returning light became visible, the southern horizon being touched for a short time with a distinct orange hue. The sun had, perhaps, afforded them a kind of illumination before, but if so, it was not to be distinguished from the "cold light of stars." They had been nearing the sunshine for thirty-two days, and had just reached that degree of mitigated darkness which made the extreme midnight of Sir Edward Parry in lat. 74° 47′.

We have already alluded to the depressing influence exercised by the prolonged and intense darkness of the Arctic night, and we have referred to the singular effect it has upon animals.

Dr. Kane's dogs, though most of them were natives of the Arctic Circle, proved unable to bear up against it. Most of them died from an anomalous form of disease, to which the absence of light would seem to have contributed as much as the extreme cold. This circumstance seems worthy of fuller notice, and we quote, therefore, Dr. Kane's observation upon it:—

"*January 20.*—This morning at five o'clock—for I am so afflicted with the *insomnium* of this eternal night, that I rise at any time between midnight and noon—I went upon deck. It was absolutely dark, the cold not permitting a swinging lamp. There was not a glimmer came to me through the ice-crusted window-panes of the cabin. While I was feeling my way, half puzzled as to the best method of steering clear of whatever might be before me, two of my Newfoundland dogs put their cold noses against my hand, and instantly commenced the most exuberant antics of satisfaction. It then occurred to me how very dreary and forlorn must these poor animals be, at atmospheres + 10° in-doors and − 50° without,—living in darkness, howling at an accidental light, as if it reminded them of the moon,—and with nothing, either of instinct or sensation, to tell them of the passing hours, or to explain the long-lost daylight."

The effect of the prolonged darkness upon these animals was most extraordinary. Every attention was paid to their wants; they were kept below, tended, fed, cleansed, caressed, and *doctored;* still they grew worse and worse. Strange to say, their disease was as clearly *mental* as in the case of any human being. There was no physical disorganization; they ate voraciously; they slept soundly, they retained their strength. But first they were stricken by epilepsy, and this was followed by true lunacy. They barked frenziedly at nothing; they walked in straight and curved lines with anxious and unwearying perseverance. They fawned on the seamen, but without seeming to appreciate any caresses bestowed upon them; pushing their head against the friend who noticed them, or oscillating with a strange pantomime of fear. Their most intelligent actions seemed of an automatic character; sometimes they clawed at their masters, as if seeking to burrow into their seal-skins; sometimes they preserved for hours a moody silence, and then started off howling, as if pursued, and ran to and fro for a considerable period.

When spring returned Dr. Kane had to mourn the loss of nine splendid Newfoundland and thirty-five Eskimo dogs; of the whole pack only six survived, and one of these was unfit for draught.

Having dwelt at some length on the characteristics of the Arctic winter, we now turn to consider those of the Arctic spring. This begins in April, but does not exhibit itself in all the freshness of its beauty until May. The temperature rises daily in the interval; the winter fall of snow, which has so long shrouded the gaunt hills and lain upon the valleys, rolls up before the rays of the rising sun; and the melted snow pours in noisy torrents and flashing cascades through the rugged ravines and over the dark sides of the lofty cliffs: everywhere the air resounds with the din of falling waters. Early in June the traveller sees with delight the signs of returning vegetation. The willow-stems grow green with the fresh and living sap; mosses, and poppies, and saxifrages, and the cochlearia, with other hardy plants, begin to sprout; the welcome whirr of wings is brought upon the breeze; the cliffs are alive with the little auks; flocks of stately eider-ducks sail into the creeks and sounds; the graceful terns scream and dart over the sea; the burgomasters and the gyrfalcons move to and fro with greater dignity; the long-tailed duck fills the echoes with its shrill voice; the snipes hover about the fresh-water pools; the sparrows chirp from rock to rock; long lines of cackling geese sail in the blue clearness overhead on their way to a remoter north; the walrus and the seal bask on the ice-floes

ADVENT OF SPRING IN THE POLAR REGIONS.

which have broken up into small rafts, and drift lazily with the currents; and a fleet of icebergs move southwards in solemn and stately procession, their spires and towers flashing and coruscating in the sunlight.

We transcribe a sketch of a spring landscape in the Polar world from the pages of Dr. Hayes:—

We arrived at the lake, he says, in the midst of a very enlivening scene. The snow had mainly disappeared from the valley, and, although no flowers had yet appeared, the early vegetation was covering the banks with green, and the feeble growths opened their little leaves almost under the very snow, and stood alive and fresh in the frozen turf, looking as glad of the spring as their more ambitious cousins of the warm South. Numerous small herds of reindeer had come down from the mountains to fatten on this newly budding life. Gushing rivulets and fantastic waterfalls mingled their pleasant music with the ceaseless hum of birds, myriads of which sat upon the rocks of the hill-side, or were perched upon the cliffs, or sailed through the air in swarms so thick that they seemed like a dark cloud passing before the sun. These birds were the little auk, a water-fowl not larger than a quail. The swift flutter of their wings and their constant cry filled the air with a roar like that of a storm advancing among the forest trees. The valley was glowing with the sunlight of the early morning, which streamed in over the glacier, and robed hill, mountain, and plain in brightness.

Spring passes into summer, and all nature seems endowed with a new life. The death-like silence, the oppressive darkness, the sense of fear and despondency, all have passed away;

and earth and water echo with cheerful voices, the landscape is bathed in a glorious radiance, the human soul is conscious of a sentiment of hope and expectation. The winter is past and gone; the flowers appear on the earth; the time of the singing of birds is come. The snow has melted from the hills, and the streams run with a merry music, and the scanty flora of the far northern world attains its full development. By day and night the sun pours forth its invigorating rays, and even the butterfly is encouraged to sport among the blossoms. The Aurora no longer exhibits its many-coloured fires, and the sky is as clear and cloudless as in genial Italy. But this season of life and warmth is of short duration, and when July has passed the sun begins to sink lower and lower, as if to visit another world; a shadow gradually steals over the sky; winds blow fiercely, and bring with them blinding showers of sleet and icicles; the fountains and the streams cease their pleasant flow; the broad crust of ice spreads over the imprisoned sea; the snow-mantle rests on the hill-sides and the valleys; the birds wing their way to the warmer South; and the Polar world is once more given over to the silence, the loneliness, and the gloom of the long Arctic night.

Turning our attention now to the "starry heavens," we observe that conspicuous among the glorious host is the North Star, which, from earliest times, has been the friend and guide of the navigator.

The Pole-Star, or Polaris, is the star α in the constellation of *Ursa Minor*, and is the nearest large star to the north pole of the celestial equator. We say the "nearest," because it does not actually mark the position of the pole, but is about 1° 30′ from it. Owing, however, to the motion of the pole of the celestial equator round that of the ecliptic, it will, in about 2000 A.D., approach within 28′ of the north pole; but after reaching this point of approximation it will begin to recede. At the time of Hipparchus it was 12° distant from it (that is, in 156 B.C.); in 1785, 2° 2′. You may easily find its place in the "stellar firmament," for a line drawn between the stars α and β (hence called the "Pointers") of the constellation *Ursa Major*, or the Great Bear, and produced in a northerly direction for about four and a half times its own length, will almost touch the Pole-Star. Two thousand years this post of honour, so to speak, was occupied by the star β of *Ursa Major*; while, in about twelve thousand years, it will be occupied by the star Vega in *Lyra*, which will be within 5° of the north pole.

The constellation of *Ursa Major* is always above the horizon of Europe, and hence it has been an object of curiosity to its inhabitants from the remotest antiquity. Our readers may easily recognize it by three stars which form a triangle in its tail, while four more form a quadrangle in the body of the imaginary bear. In the triangle, the first star at the tip of the tail is Benetnasch of the second magnitude; the second, Mizar; and the third, Alioth. In the quadrangle, the first star at the root of the tail is named Megrez; the second below it, Phad; the third, in a horizontal direction, Merak; and the fourth, above the latter, Dubhe, of the first magnitude.

URSA MAJOR AND URSA MINOR.

In *Ursa Minor* the only conspicuous star is Polaris, of which we have recently spoken.

We subjoin a list of the northern constellations, including the names of those who formed them, the number of their visible stars, and the names of the most important and conspicuous.

NORTHERN CONSTELLATIONS.

Constellations.	Author.	No of Stars.	Principal Stars.
Ursa Minor, the Lesser Bear	Aratus.	24	Polaris, 2.
Ursa Major, the Great Bear	Aratus.	87	Dubhe, 1; Alioth, 2.
Perseus, and Head of Medusa	Aratus.	59	Algenib, 2; Algol, 2.
Auriga, the Waggoner	Aratus.	66	Capella, 1.
Boötes, the Herdsman	Aratus.	54	Arcturus, 1.
Draco, the Dragon	Aratus.	80	Rastaben, 3.
Cepheus	Aratus.	35	Alderamin, 3.
Canes Venatici, the Greyhounds Chara and Asteria	Hevelius.	25	......
Cor Caroli, Heart of Charles II	Halley.	3	......
Triangulum, the Triangle	Aratus.	16	......
Triangulum Minus, the Lesser Triangle	Hevelius.	10	......
Musca, the Fly	Bode.	6	......
Lynx	Hevelius.	44	......
Leo Minor, the Lesser Lion	Hevelius.	53	......
Coma Berenices, Berenice's Hair	Tycho Brahe.	43	......
Cameleopardalis, the Giraffe	Hevelius.	58	......
Mons Menelaus, Mount Menelaus	Hevelius.	11	......
Corona Borealis, the Northern Crown	Aratus.	21	......
Serpens, the Serpent	Aratus.	64	......
Scutum Sobieski, Sobieski's Shield	Hevelius.	8	......
Hercules, with Cerberus	Aratus.	113	Ras Algratha, 3.
Serpentarius, or Ophiuchus, the Serpent-Bearer	Aratus.	74	Ras Aliagus, 2.
Taurus Poniatowski, or the Bull of Poniatowski	Poezobat.	7	......
Lyra, the Harp	Aratus.	22	Vega, 1.
Vulpeculus et Anser, the Fox and the Goose	Hevelius.	37	......
Sagitta, the Arrow	Aratus.	18	......
Aquila, the Eagle, with Antinous	Aratus.	71	Altair, 1.
Delphinus, the Dolphin	Aratus.	18	......
Cygnus, the Swan	Aratus.	81	Deneb, 1.
Cassiopeia, the Lady in her Chair	Aratus.	55	......
Equulus, the Horse's Head	Ptolemy.	10	......
Lacerta, the Lizard	Hevelius.	10	......
Pegasus, the Flying Horse	Aratus.	89	Markab, 2.
Andromeda	Aratus.	66	Almaac, 2.
Turandus, the Reindeer	Lemonnier.	12	......

A few remarks in reference to some of these constellations, and the glorious orbs which they help to indicate to mortal eyes, may fitly close this chapter.

We have already alluded to *Ursa Major*, which forms one of the most conspicuous objects of the northern heavens. It has borne different names, at different times, and among different peoples. It was the Ἄρκτος μεγάλη of the Greeks; the "Septem triones" of the Latins. It is known in some parts as David's Chariot; the Chinese call it, *Tcheou-pey*.

Night and day this constellation watches above the northern horizon, revolving, with slow and majestic march, around Polaris, in four and twenty hours. The quadrangle of stars in the body of the Great Bear forms the wheels of the chariot; the triangle in its tail, the chariot-pole. Above the second of the three latter shines the small star Alcor, also named the Horseman. The Arabs call it Saidak, or "the Test," because they use it to try the range and strength of a person's vision.

This brilliant northern constellation, composed, with the exception of δ, of stars of the second magnitude, has frequently been celebrated by poets. We may paraphrase, for the advantage of our readers, a glowing apostrophe from the pen of the American Ware:—

With what grand and majestic steps, he says, it moves forward in its eternal circle, following among the stars its regal way in a slow and silent splendour! Mighty creation, I salute thee!

I love to see thee wandering in the shining paths like a giant proud of his strong girdle—severe, indefatigable, resolved—whose feet never lag in the road which lies before them. Other tribes abandon their nocturnal course and rest their weary orbs under the waves; but thou, thou never closest thy burning eyes, and never suspendest thy determined steps. Forward, ever forward! While systems change, and suns retire, and worlds fall to sleep and awake again, thou pursuest thy endless march. The near horizon attempts to check thee, but in vain. A watchful sentinel, thou never quittest thy age-long duty; but, without allowing thyself to be surprised by sleep, thou guardest the fixed light of the universe, and preventest the north from ever forgetting its place.

Seven stars dwell in that shining company; the eye embraces them all at a single glance; their distances from one another, however, are not less than the distance of each from Earth. And this again is the reciprocal distance of the celestial centres or foci. From depths of heaven, unexplored by thought, the piercing rays dart across the void, revealing to our senses innumerable worlds and systems. Let us arm our vision with the telescope, and let us survey the firmament. The skies open wide; a shower of sparkling fires descends upon our head; the stars close up their ranks, are condensed in regions so remote that their swift rays (swifter than aught else in creation) must travel for centuries before they can reach our Earth. Earth, sun, and ye constellations, what are ye among this infinite immensity and the multitude of the Divine works!

If we face towards the Pole-Star, which, as we have seen, preserves its place in the centre of the northern region of the sky, we have the south behind us, the east is on our right, the west upon our left. All the stars revolving round the Pole-Star, from right to left, should be recognized according to their mutual relations rather than referred to the cardinal points. On the other side of Polaris, as compared with the Great Bear, we find another constellation which is easily recognized. If from the central star δ we carry a line to the Pole, and then prolong it for an equal distance, we traverse the constellation of *Cassiopeia*, composed of five stars of the third magnitude, disposed somewhat like the outer jambs of the letter M. The small star χ, terminating the square, gives it also the form of a chair. This group occupies every possible situation in revolving round the Pole, being at one time above it, at another below, now on the left, and then on the right; but it is always readily found, because, like *Ursa Major*, to which it is invariably opposite, it never sets. The Pole-Star is the axle round which these two constellations revolve.

If we now draw, from the stars α and δ in *Ursa Major*, two lines meeting at the Pole, and afterwards extend them beyond *Cassiopeia*, they will abut on the square of *Pegasus*, which is bounded on one of its sides by a group, or series, of three stars resembling the triangle in *Ursa Major*. These three belong to the constellation of *Andromeda* (α, β, and γ), and themselves abut on another three-orbed group, that of *Perseus*.

The last star in the square of *Perseus* is also the first α of Andromeda: the other three are named, Algenib, γ; Markab, α; and Scheat, β. To the north of Andromeda β, and near a small star, ν, the Arctic traveller will discern an oblong nebula, which may be compared to the light of a taper seen through a sheet of horn; this is the first nebula to which any allusion occurs in the annals of astronomy. In *Perseus* α, an orb of great brilliancy, on the prolonged plane of the three principal stars of *Andromeda*, shines with steady lustre between two less dazzling spheres, and forms in conjunction with them a concave arc very easily distinguished. Of this arc we may avail ourselves as a new point of departure. By prolonging it in the direction of δ, we come to a very bright star of the first magnitude, the *Goat*. By forming a right angle to this prolongation in a southerly direction we come to that glorious mass of stars, not very frequently above the Polar horizon, the *Pleiads*. These were held in evil repute among the ancients. Their appearance was supposed to be ominous of violent storms, and Valerius Flaccus speaks of them as fatal to ships.

NEBULA IN ANDROMEDA.

Algol, or Medusa's Head, known to astronomers as Perseus β, belongs to the singular class of Variable Stars. Instead of shining with a constant lustre, like other orbs, it is sometimes very brilliant, and sometimes very pale; passing, apparently, from the second to the fourth magnitude. According to Goodricke, its period of variation is 2 days 20 hours 48 minutes. This phenomenal character was first observed by Maraldi in 1694; but the duration of the change was determined by Goodricke in 1782. For two days and fourteen hours it continues at its brightest, and shines a glory in the heavens. Then its lustre suddenly begins to wane, and in three hours and a half is reduced to its minimum. Its weakest period, however, does not last more than about fifteen minutes. It then begins to increase in brightness, and in three hours and a half more it is restored to its full splendour; thus passing through its succession of changes in 2 days 20 hours 48 minutes.

This singular periodicity suggested to Goodricke the idea of some opaque body revolving around the star, and by interposing between it and the Earth cutting off a portion of its light. Algol is one of the most interesting of the welcome stars which kindle in the long Arctic darkness.

The star ζ in *Perseus*, situated above the "stormy Pleiads," is double; that is, a binary star. ξ in Ursa Major is also a twin-star; and so is Polaris, the second and smaller star appearing a mere speck in comparison with its companion.

These are the principal stars and starry groups in the Circumpolar Regions of the heavens, on one side; let us now turn our attention to the other.

For this purpose we must again take the Great Bear as our starting-point. Prolonging the tail in its curvature, the Arctic traveller notes, at some distance from it, a star of the first magnitude, Arcturus, or Boötes α. This star, though without any authority, was at one time considered the nearest to the Earth of all the starry host. About 10° to the north-east of it is Mirac, or ϵ Boötes; one of the most beautiful objects in the heavens, on account of the contrasted hues, yellow and azure, of the two stars composing it. Unfortunately, the twin-orbs cannot be distinctly seen except with a telescope of two hundred magnifying power.

A small ring of stars to the left of Boötes is appropriately known as *Corona Borealis*, or the Northern Crown.

The constellation of *Boötes* forms a pentagon; and the stars composing it are all of the third magnitude, with the exception of α, which is of the first. Arcturus, as we have said, was anciently considered the star nearest to the Earth. It is, at all events, *one* of the nearest, and belongs to the small number of those whose distance our astronomers have succeeded in calculating. It is 61 trillions, 712,000 millions of leagues from our planet; a distance of which we can form no appreciable conception. Moreover, it is a coloured star; on examining it through a telescope we see that it is of the same hue as the "red planet Mars."

By carrying a line from the Polar Star to Arcturus, and raising a perpendicular in the middle of this line, opposite to *Ursa Major*, the observer of the Arctic skies will discover one of the most luminous orbs of night, *Vega*, or α Lyra, near the Milky Way. The star β Lyra, or Sheliak, is a variable star, changing from the third to the fifth magnitude, and accomplishing its variation in 6 days 10 hours and 34 minutes. β and ε Lyra are quadruple systems, each composed of binary or twin-stars.

The line drawn from Arcturus to Vega cuts the constellation of Hercules.

Between Ursa Major and Ursa Minor may be observed a prolonged series of small stars, coiling, as it were, in a number of convolutions, and extending towards Vega: these belong to the constellation of the *Dragon*.

Such are the principal objects which attract the attention of the traveller, when contemplating the star-studded firmament of the Arctic night.

CHAPTER III.

THE POLAR SEAS: ICEBERGS—ICE-FLOES—THE SEAL—THE WALRUS—THE NARWHAL—THE WHALE—SUNDRY FORMS OF MARINE LIFE.

THOSE masses of ice which, towering to a considerable elevation above the surface of the water, are carried hither and thither by the currents of the Polar Sea, are known as *Icebergs*. They are fresh-water formations, originating in the great glaciers of the northern highlands. For as the rivers continuously pour their waters into the ocean, so do the glaciers incessantly glide downward from the head of the valleys which they occupy, until, arriving on the coast, they throw off their terminal projections, to be carried afar by the action of the tidal waves.

These bergs, or floating mountains, are sometimes 250 to 300 feet above the level of the sea, and their capacity or bulk is invariably equal to their height. From their specific gravity it has been calculated that the volume of an iceberg *below* the water is eight times that of the portion rising above it. They are frequently of the most imposing magnitude. Ross, in his first expedition, fell in with one in Baffin Bay, at a distance of seven leagues from land, which had gone aground in sixty-one fathoms water. Its dimensions, according to Lieutenant Parry, were 4,169 yards in length, 3,869 yards in breadth, and 51 feet in height. Its configuration is described as resembling that of the back of the Isle of Wight, while its cliffs recalled those chalky ramparts which stretch their glittering line to the west of Dover. Its weight was computed at 1,292,397,673 tons. Captain Graab examined a mass, on the west coast of Greenland, which rose 120 feet out of the water, measured 4,000 feet in circumference at the base, and was calculated to be equal in bulk to upwards of 900,000,000 cubic feet. Dr. Hayes took the measurements of a berg which had stranded off the little harbour of Tessuissak, to the north of Melville Bay. The square wall which faced towards his base of triangulation was somewhat more than three-quarters of a mile in length, and 315 feet in height. As it was nearly square-sided above the sea, it would be of the same shape beneath it; and, according to the ratio already given, must have drifted aground in a depth of fully half a mile. In other words, from base to summit it must have stood as high as the peak of Snowdon. Its cubical contents cannot have been less than about 27,000,000,000 feet, nor its weight than 2,000,000,000 tons!

When seen from a distance, the spectacle of any considerable number of these slowly-moving mountains is very impressive, and it becomes particularly magnificent if it should be lit up by the splendour of the midnight sun. They are not only majestic in size, but sublime in appearance, at one time assuming the likeness of a grand cathedral church, at another, of a lofty obelisk; now of a dazzling pyramid, and now of a cluster of lofty towers. Nature would seem to have

ARCHED ICEBERG OFF THE GREENLAND COAST.

lavished upon them all her architectural fancy; and as they are grandly swept along, one might be pardoned for supposing them to be the sea-washed palaces of a race of ocean Titans.

In Melville Bay, Dr. Kane's ship anchored to an iceberg, which protected it from the fury of a violent gale. But he had not long enjoyed the tranquil shelter it afforded, when a din of loud crackling sounds was heard above; and small fragments of ice, not larger than a walnut, began to dot the water, like the first big drops of a thunder-shower. Dr. Kane and his crew did not neglect these indications; they had barely time to cast off, however, before the face of the icy cliff fell in ruins, crashing like near artillery.

Afterwards he made fast to a larger berg, which he describes as a moving breakwater, and of gigantic proportions; it kept its course steadily towards the north.

When he got under weigh, and made for the north-east, through a labyrinth of ice-floes, he was favoured with a gorgeous spectacle, which hardly any excitement of peril could have induced him to overlook. The midnight sun came out over the northern crest of the huge berg, kindling variously-coloured fires on every part of its surface, and making the ice around one sublime transparency of illuminated gem-work, blazing carbuncles, and rubies and molten gold.

Dr. Hayes describes an immense berg which resembled in its general aspect the Westminster Palace of Sir Charles Barry's creation. It went to ruin before his eyes. First one tall tower tumbled headlong into the water, starting from its surface an innumerable swarm of gulls; then another followed; and at length, after five hours of terrible disruption and crashing, not a fragment that rose fifty feet above the water remained of this architectural colossus of ice.

AMONG THE BERGS—A NARROW ESCAPE.

These floating isles of ice are carried southward fully two thousand miles from their parent glaciers to melt in the Atlantic, where they communicate a perceptible coldness to the water for thirty or forty miles around, while their influence on the atmospheric temperature may be recognized at a greater distance. Their number is extraordinary. As many as seven hundred bergs, each loftier than the dome of St. Paul's, some than the cross of St. Peter's, have been seen at once in the Polar basin; as if the Frost King had despatched an armada to oppose the rash enterprise of man in penetrating within his dominions. The waves break against them as against an iron-bound coast, and often the spray is flung over their very summits, like the spray of the rolling waters of the Channel over the crest of the Eddystone Lighthouse. The ice crumbles from their face, and tumbles down into the sea with a roar like that of artillery; and as they waste away, through the combined action of air and water, they occasionally lose their equilibrium and topple over, producing a swell and a violent commotion which break up the neighbouring ice-fields: the tumult spreads far and wide, and thunder seems to peal around.

The fractures or rents frequently visible in the glittering cliffs of the icebergs are of an emerald green, and look like patches of beautiful fresh sward on cliffs of chalk; while pools of water of the most exquisite sapphirine blue shine resplendent on their surface, or leap down their craggy sides in luminous cascades. Even in the night they are readily distinguished from afar by their effulgence; and in foggy, hazy weather, by a peculiar blackness in the atmosphere. As the Greenland Current frequently drifts them to the south of Newfoundland, and even to the 40th or 39th parallel of latitude, the ships and steamers crossing between Europe and America sometimes meet them on their track. To come into collision with them is certain destruction; and it is probable that some of those ill-fated vessels which have left their harbours in safety, but have never since been heard of,—as, for example, the steamer *President*,—have perished through this cause.

But if they are sometimes dangerous to the mariner, they often prove his security. As most of their bulk lies below the water-surface, they are either carried along by under-currents against the wind, or else from their colossal size they are able to defy the strongest gale, and to move along with majestic slowness when every other kind of ice is driven swiftly past them. And hence it happens that, when the wind is contrary, the whaler is glad to bring his ship into smooth water under their lee. In describing the difficulties of his passage through the loose and drifting ice near Cape York, and the broken ice-fields, Dr. Kane records the assistance he derived from the large icebergs, to which he moored his vessel, and thus was enabled, he says, to hold his own, however rapidly the surface-floes were passing by him to the south.

Yet anchoring to a berg brings with it an occasional peril. As we have already said, large pieces frequently loosen themselves from the summit or sides, and fall into the sea with a far-resounding crash. When this operation, "calving," as it is called, takes place, woe to the unfortunate ship which lies beneath!

All ice becomes excessively brittle under the influence of the sun or of a temperate atmosphere, and a single blow from an axe will suffice to split a huge berg asunder, burying the heedless adventurer beneath the ruins, or hurling him into the yawning chasm.

Dr. Scoresby records the adventure of two sailors who had been sent to attach an anchor to a berg. They set to work to hew a hole in the ice, but scarcely had the first blow been struck, when the colossal mass rent from top to bottom and fell asunder, the two halves falling in opposite directions with a tremendous uproar. One of the sailors, with remarkable presence of mind, instantly clambered up the huge fragment on which he was sitting, and remained rocking to and fro on the dizzy summit until its equilibrium was restored; the other, falling between the masses, would probably have been crushed to death if the current caused by their commotion had not swept him within reach of the boat that was waiting for them.

Fastening to a berg, says Sherard Osborn, has its risks and dangers. Sometimes the first stroke of the man setting the ice-anchor, by its concussion, causes the iceberg to break up, and the people so employed run great risk of being injured; at another time, vessels obliged to make fast under the steep side of a berg have been seriously damaged by pieces detaching themselves from overhead; and, again, the projecting masses, called tongues, which form under water the base of the berg, have been known to break off, and strike a vessel so severely as to sink her. All these perils are duly detailed by every Arctic navigator, who is always mindful, in mooring to an iceberg, to look for a side which is low and sloping, without any tongues under water.

Captain Parry was once witness of that sublime spectacle, which, though of frequent occurrence, is seldom seen by human eyes, the entire dissolution of an enormous iceberg.

Its huge size and massiveness had been specially remarked, and men thought that it might well resist "a century of sun and thaw." It looked as large as Westminster Abbey. All on board Captain Parry's ship described as a most wonderful spectacle this iceberg, without any warning, completely breaking up. The sea around it became a seething caldron, from the violent plunging of the masses, as they broke and re-broke in a thousand pieces. The floes, torn up for a distance of two miles around it, by the violent action of the rolling waters, threatened, from the agitation of the ice, to destroy any vessel that had been amongst them; and Captain Parry and his crew congratulated themselves that they were sufficiently far from the scene to witness its sublimity without being involved in its danger.

Icebergs chiefly abound in Baffin Bay, and in the gulfs and inlets connected with it. They are particularly numerous in the great indentation known as Melville Bay, the whole interior of the country bordering upon it being the seat of immense glaciers, and these are constantly "shedding off" icebergs of the largest dimensions. The greater bulk of these is, as we have explained, below the water-line; and the consequent depth to which they sink when floating

ICEBERG AND ICEFIELD, MELVILLE BAY, GREENLAND

subjects them to the action of the deeper ocean-currents, while their broad surface above the water is, of course, acted on by the wind. It happens, therefore, as Dr. Kane remarks, that they are found not infrequently moving in different directions from the floes around them, and preventing them for a time from freezing into a united mass. Still, in the late winter, when the cold has thoroughly set in, Melville Bay becomes a continuous mass of ice, from Cape York to the Devil's Thumb. At other times, this region justifies the name the whalers have bestowed upon it of "Bergy Hole."

Captain Beechey, in his voyage with Buchan, in 1818, had an opportunity of witnessing the formation of a "berg," or rather of two of these immense masses. In Magdalena Bay he had taken the ship's launch near the shore to examine a magnificent glacier, when the discharge of a gun caused an instantaneous disruption of its bulk. A noise resembling thunder was heard in the direction of the glacier, and in a few seconds more an immense piece broke away, and fell headlong into the sea. The crew of the launch, supposing themselves beyond the reach of its influence, quietly looked upon the scene, when a sea arose and rolled towards the shore with such rapidity that the boat was washed upon the beach and filled. As soon as their astonishment had subsided, they examined the boat, and found her so badly stove that it was necessary to repair her before they could return to their ship. They had also the curiosity to measure the distance the boat had been carried by the wave, and ascertained that it was ninety-six feet.

A short time afterwards, when Captain Beechey and Lieutenant Franklin had approached one of these stupendous walls of ice, and were endeavouring to search into the innermost recess of a deep cavern that lay near the foot of the glacier, they suddenly heard a report, as of a cannon, and turning to the quarter whence it proceeded, perceived an immense section of the front of the glacier sliding down from the height of two hundred feet at least into the sea, and dispersing the water in every direction, accompanied by a loud grinding noise, and followed by an outflow of water, which, being previously lodged in the fissures, now made its escape in innumerable tiny flashing rills and cataracts.

The mass thus disengaged at first disappeared wholly under water, and nothing could be seen but a violent seething of the sea, and the ascent of clouds of glittering spray, such as that which occurs at the foot of a great waterfall. But after a short time it re-appeared, raising its head fully a hundred feet above the surface, with water streaming down on every side; and then labouring, as if doubtful which way it should fall, it rolled over, rocked to and fro for a few minutes, and finally became settled.

On approaching and measuring it, Beechey found it to be nearly a quarter of a mile in circumference, and sixty feet out of the water. Knowing its specific gravity, and making a fair allowance for its inequalities, he computed its weight at 421,660 tons.

In Parry's first voyage he passed in one day fifty icebergs of large dimensions, just after crossing the Arctic Circle; and on the following day a still more extended chain of ice-peaks of still larger size, against which a heavy southerly swell was violently driven, dashing the loose ice with tremendous force, sometimes flinging a white spray over them to the height of more than one hundred feet, and accompanied by a loud noise "exactly resembling the roar of distant thunder."

Between one of these bergs and a detached floe the *Hecla*, Parry's ship, had nearly, as the whalers say, been "nipped," or crushed. The berg was about one hundred and forty feet high, and aground in one hundred and twenty fathoms, so that its whole height must have exceeded eight hundred feet; that is, it was of a bulk equal to St. Catherine's Down in the Isle of Wight.

In his second voyage Parry speaks of fifty-four icebergs visible at one time, some of which were not less than two hundred feet above the sea; and again of thirty of these huge masses, many of them whirled about by the tides like straws on a mill-stream.

Icebergs can originate only in regions where glaciers abound: the former are the offspring of the latter, and where land unsuitable to the production of the latter does not exist, the former are never found. Hence, in Baffin Bay, where steep cliffs of cold granite frown over almost fathomless waters, the "monarch of glacial formations" floats slowly from the ravine which has been its birth-place, until fairly launched into the depths of ocean, and, "after long years," drifts into the warmer regions of the Atlantic to assist in the preservation of Nature's laws of equilibrium of temperature of the air and water.

There was a time when men of science, and, amongst others, the French philosopher St. Pierre, believed that icebergs were the snow and ice of ages accumulated upon an Arctic sea, which, forming at the Poles, detached themselves from the parent mass. Such an hypothesis naturally gave rise to many theories, not less ingenious than startling, as to the effect an incessant accumulation of ice must produce on the globe itself; and St. Pierre hinted at the possibility of the huge "domes of ice"—which, as he supposed, rose to an immense height in the keen frosty heavens of the Poles—suddenly launching towards the Equator, dissolving under a tropical sun, and resulting in a second deluge!

In simple language Professor Tyndall furnishes an explanation of the origin of icebergs, which we may transfer to these pages as supplementary to the preceding remarks.

What is their origin? he asks; and he replies, as we have done, the Arctic glaciers. From the mountains in the interior the indurated snows slide into the valleys, and fill them with ice. The glaciers thus created move, like the Swiss ones, incessantly downward. But the Arctic glaciers descend to the sea, and even enter it, frequently ploughing up its bottom into submarine moraines. Undermined by the continuous action of the waves, and unable to resist the pressure of their own weight, they break across, and discharge enormous masses into the ocean. Some of these drift on the adjacent shores, and often maintain themselves for years. Others float away to the southward, and pass into the broad Atlantic, where they are finally dissolved. But a vast amount of heat is demanded for the simple liquefaction of ice, and the melting of icebergs is on this account so slow that, when large, they sometimes maintain themselves till they have been drifted two thousand miles from their place of birth.

ORIGIN OF ICEBERGS—EXTENSION OF A GLACIER SEAWARDS.

Icebergs, then, are fresh-water formations, and though they are found on a colossal scale only in the Polar seas, yet they are by no means uncommon among the lofty Alpine lakes.

The monarch of European ice-rivers is the great Aletsch glacier, at the head of the valley of the Rhone. It is about twenty miles in length, and collects its materials from the snow-drifts of the grandest mountains of the Bernese Oberland—the Jungfrau, the Mönch, the Trugberg, the Aletschhorn, the Breithorn, and the Gletscherhorn.

From the peak of the Æggischhorn the Alpine traveller obtains a fine view of its river-like course, and he sees beneath him, on the right hand, and surrounded by sheltering mountains, an object of almost startling beauty. "Yonder," says Tyndall,* "we see the naked side of the glacier, exposing glistening ice-cliffs sixty or seventy feet high. It would seem as if the Aletsch

* Tyndall, "Forms of Water," p. 137.

THE ALETSCH GLACIER, SWITZERLAND, FROM THE ÆGGISCHHORN, SHOWING ITS MORAINES.

here were engaged in the vain attempt to thrust an arm through a lateral valley. It once did so; but the arm is now incessantly broken off close to the body of the glacier, a great space formerly covered by the ice being occupied by its water of liquefaction. In this way a lake of the loveliest blue is formed, which reaches quite to the base of the ice-cliffs, saps them, as the Arctic waves sap the Greenland glaciers, and receives from them the broken masses which it has undermined. As we look down upon the lake, small icebergs sail over the tranquil surface, each resembling a snowy swan accompanied by its shadow."

THE MARJELEN SEA, SWITZERLAND.

This lake is the Märjelen Sea of the Swiss.

Professor Tyndall goes on to describe a spectacle which he witnessed, and which, as we have seen, is of frequent occurrence in the Arctic Seas. A large and lonely iceberg was floating in the middle of the lake. Suddenly he heard a sound like that of a cataract, and on looking towards the iceberg could see the water teeming from its sides. Whence came the water? The berg had become top heavy through the melting underneath; it was in the act of performing a somersault, and in rolling over carried with it a vast quantity of water, which rushed like a waterfall down its sides. And the iceberg, which, but a moment before, was snowy white, now exhibited the delicate blue colour characteristic of compact ice. It would soon, however, be rendered white again by the action of the sun.

We may contrast this picture of the solitary iceberg in the centre of the dark-blue lake with one which Dr. Hayes describes in his picturesque voyage in the open Polar Sea.

After passing Upernavik he saw a heavy line of icebergs lying across his course, and having no alternative, shot in among them. Some of them proved to be of immense size—upwards of two hundred feet in height, and a mile in length; others were not larger than the schooner which wound her way amongst them. Their forms were as various as their dimensions, from solid wall-sided masses of dead whiteness, with waterfalls tumbling from them, to an old weather-worn accumulation of Gothic spires, whose crystal peaks and sharp angles melted into the blue sky. They seemed to be endless and innumerable, and so close together that at a little distance they appeared to form upon the sea an unbroken canopy of ice.

Dr. Hayes records an adventure which may serve to give the reader an idea of the nature of the perils encountered by the Arctic explorer. The ocean-current was carrying his schooner towards a labyrinth of icebergs at an uncomfortably rapid rate. A boat was therefore lowered, to moor a cable to a berg which lay grounded at about a hundred yards distant. While this was being done the schooner absolutely grazed the side of a berg which rose a hundred feet above her topmasts, and then slipped past another of smaller dimensions. But a strong eddy at this moment carried her against a huge floating mass, and though the shock was slight, it proved sufficient to disengage some fragments of ice large enough to have crushed the vessel had they struck her. The berg then began to revolve, slowly and ponderously, and to settle slowly over the threatened ship, whose destruction seemed a thing of certainty.

Fortunately, she was saved by the action of the berg. An immense mass broke off from that part which lay beneath the water-surface, and this colossal fragment, a dozen times larger than the schooner, came rushing up within a few yards of them, sending a vast volume of foam and water flying from its sides. This rupture arrested the rotatory motion of the berg, which then began to settle in another direction, and the schooner was able to sheer off.

At this moment the crew were startled by a loud report. Another and another followed in quick succession, until the din grew deafening, and the whole air seemed a reservoir of chaotic sounds. The opposite side of the berg had split off, piece after piece, toppling a vast volume of ice into the sea, and sending the berg revolving back upon the ship. Then the side nearest to them underwent the same singular process of disruption, and came plunging wildly down into the sea, sending over them a shower of spray, and raising a swell which rocked the ship to and fro as in a gale of wind, and left her grinding in the *débris* of the crumbling ruin.

"The ice was here,
The ice was there,
The ice was all around;
It creaked and growled,
And roared and howled,
Like demons in a swound."

It is impossible, we should say, for any one who has not had actual experience of the conditions of the Arctic world, to comprehend or imagine the immense quantity of ice upborne on its cold bleak waters. The mere enumeration of the floating bergs at times defies the navigator. Dr. Hayes once counted as far as five hundred, and then gave up in despair. Near by they stood out, he says, in all the rugged harshness of their sharp outlines; and from this, softening with the distance, they melted away into the clear gray sky; and there, far off upon the sea of liquid silver, the imagination conjured up the strangest and most wonderful groups and objects. Birds and beasts and human forms and architectural designs took shape in the distant masses of blue and white. The dome of St. Peter's was recognizable here; then the spire of a village church rose sharp and distinct; and under the shadow of the Pyramids nestled a Byzantine tower and a Grecian temple.

"To the eastward," says Dr. Hayes, describing a similar scene, "the sea was dotted with little islets—dark specks upon a brilliant surface. Icebergs, great and small, crowded through the channels which divided them, until in the far distance they appeared massed together, terminating against a snow-covered plain that sloped upward until it was lost in a dim line of bluish whiteness. This line could be traced behind the serrated coast as far to the north and south as the eye could carry. It was the great *Mer de Glace** which covers the length and breadth of the Greenland continent. The snow-covered slope was a glacier descending therefrom—the parent stem from which had been discharged, at irregular intervals, many of the icebergs which troubled us so much."

We have now brought together a sufficient number of data to assist the reader in forming a vivid conception of those monsters of the Polar Seas, the icebergs; and to enable him, unless he is very slow of imagination, to realize to himself what they *are*, and what their general aspect *is*. But we may add one interesting detail, noticed by Mr. Lamont, the persevering seal-hunter, which is very generally overlooked.

In the course of the brief Arctic summer the increased solar warmth has a perceptible effect upon the solid ice, and it becomes undermined and honeycombed, or, as the sailors call it, "rotten," like a chalk cliff. It decays fastest, apparently, "between wind and water," so that enormous caverns are excavated in the sides of the bergs.

Poets never dreamed of anything more beautiful than these crystal vaults, which sometimes appear of a deep ultramarine blue, and at others of an emerald-green tint. One could fancy them the favourite haunts of mermaids and mermen, and of every kind of sea monster; but, in truth, no animal ever enters them; the water dashing in and out through their icy caves and tunnels makes a sonorous but rather monotonous and melancholy sound In moderately calm weather many of these excavated bergs assume the form of gigantic mushrooms, and all kinds of fantastic outlines; but as soon as a breeze of wind arises they break up into little pieces with great rapidity.

Icebergs are met with on every side of the Southern Pole, and on every meridian of the great Antarctic Ocean. But such is not the case in the North. In the 360th meridian of longitude which intersects the parallel of 70° N., icebergs spread over an extent only of about fifty-five degrees, and this is immediately in and about Greenland and Baffin Bay. Or, as Admiral Osborn puts it, for 1,375 miles of longitude we have icebergs, and then for 7,635 geographical miles none are met with. This fact is, as the same writer calls it, most interesting, and points strongly to the probability that no extensive area of land exists about the North Pole; a sup-

* The name given to a plain of ice near Mont Blanc.

FALL OF AN ICEBERG

position strengthened by another fact, that the vast ice-fields off Spitzbergen show no signs of ever having been in contact with land or gravel.

Another difficulty which besets the Arctic navigator is the "pack-ice."

In winter, the ice from the North Pole descends so far south as to render the coast of Newfoundland inaccessible; it envelops Greenland, sometimes even Iceland, and always surrounds and blocks up Spitzbergen and Novaia Zemlaia. But as the sun comes north this vast frozen expanse, which stretches over several thousands of square miles, breaks up into enormous masses. When these extend horizontally for a considerable distance they are called *ice-fields*.

IN AN ICE-PACK, MELVILLE BAY.

A *floe* is a detached portion of a field; a large area of floes, closely compact together, is known as *pack-ice;* while *drift-ice* is loose ice in motion, and not so firmly welded as to prevent a ship from forcing her way through the yielding fragments.

This "pack-ice," however, is the great obstacle to Arctic exploration; and frequently it presents a barrier which no human enterprise or skill can overpass. At times, it has been found possible to cut a channel through it, or it breaks up and opens a water-way through which the bold adventurer steers. In 1806, Captain Scoresby forced his ship through two hundred and fifty miles of pack-ice, in imminent peril, until he reached the parallel of 81° 50′,—his nearest approach to the Pole. In 1827, Sir Edward Parry gained the latitude of 82° 45′, by dragging a boat over the ice-fields, but was then compelled to abandon his daring and hazardous attempt, because the current carried the ice southward more rapidly than he could traverse it to the north.

In warm summers this mass of ice will suddenly clear away and leave an open streak of silver sea along the west coast of Spitzbergen, varying in width from sixty to one hundred and fifty miles, and reaching as high as 80° or 80° 30′ N. latitude. It was through this channel that Scoresby bore his ship on the expedition to which we have just alluded. A direct course from the Thames,

CHANNEL IN AN ICE-FIELD.

across the Pole, to Behring Strait is 3,570 geographical miles; by Lancaster Sound it is 4,660 miles. The Russians would be saved a voyage of 18,000 geographical miles could they strike across the Pole and through Behring Strait to British Columbia, instead of going by Cape Horn.

Ice-fields, twenty to thirty miles across, are of frequent occurrence in the great Northern Ocean; sometimes they extend fully one hundred miles, so closely and solidly packed that no opening, even for a boat, intervenes between them; they vary in thickness from ten to forty or even fifty feet. At times these fields, which are many thousand millions of tons in weight, acquire a rapid rotatory motion, and dash against one another with a fury of which no words can give an accurate idea. The reader knows what awful results are produced by the collision of two railway trains, and may succeed, perhaps, in forming some feeble conception of this still more appalling scene when he remembers the huge dimensions and solidity of the opposing forces. The waters seethe and foam, as if lashed by a tremendous tempest; the air is smitten into stillness by the chaos of sounds, the creaking, and rending, and cracking, and heaving, as the two ice-fields are hurled against each other.

"NIPPED" IN AN ICE-FIELD.

Woe to the ship caught between these grinding

masses! No vessel ever built by human hands could resist their pressure; and many a whaler, navigating amid the floating fields, especially in foggy weather, has thus been doomed to destruction. Some have been caught up like reeds, and flung helplessly upon the ice; others have been overrun by the ice, and buried beneath the accumulated fragments; others have been dashed to pieces, and have gone down suddenly with all on board.

The records of Arctic exploration are full of stories of "hairbreadth escapes" from the perils of the ice-field and the ice-floe. Here is one which we borrow from the voyage of the *Dorothea* and the *Trent*, under Captain Buchan and Lieutenant Franklin.

The two vessels were making for Magdalena Bay, when they were caught in a violent storm, and compelled to heave-to under storm stay-sails. Next morning (June 30) the ice was seen along the lee, with a furious sea breaking upon it. Close-reefed sails were out in the hope of weathering the danger. When Buchan found that this could not be effected by his ship, a slow and heavy sailer, he resolved on the desperate expedient of "taking the pack," in preference to falling, broadside on, among the roaring breakers and crashing ice. "Heaven help them!" was the involuntary cry of those on board the *Trent*, and the prayer was all the more earnest from the conviction that a similar fate would soon be their own.

The *Dorothea* wore, and, impelled by wind and sea, rushed towards what seemed inevitable destruction; those in the *Trent* held their breath while they watched the perilous exploit. The suspense lasted but a moment, for the vessel, like a snow-flake before the storm, drove into the awful scene of foam, and spray, and broken ice, which formed a wall impenetrable to mortal eyesight. Whether she was lost or saved, the gallant hearts on board the *Trent* would never know until they too were forced into a manœuvre which appeared like rushing into the jaws of death. But it was inevitable; and when Franklin had made all his preparations, he gave, in firm, decisive tones, the order to "put up the helm."

No language, says Admiral Beechey, who was then serving as a lieutenant on board the *Trent*, can convey an adequate idea of the terrific grandeur of the effects produced by the collision of the ice and the tempestuous ocean. No language, on the other hand, can convey an idea of the heroic calmness and resolution of Franklin and his crew. As they approached the terrible scene, Franklin watched for one opening less hazardous than another; but there was none. Before them stretched one long line of frightful breakers, immense blocks of ice heaving, rearing, and hurtling against one another with a din which rendered the loud voice of the gallant commander almost inaudible. On the crest of a huge billow the little *Trent* rushed into the horrible turmoil; a shock, which quivered through the ship from stem to stern, and the crew were flung upon the deck, and the masts bent like willow wands.

"Hold on, for your lives, and stand to the helm, lads!" shouted Franklin. "Ay, ay, sir," was the steady response from many a heroic heart. A billow came thundering against the stern of the brig; would the brig be engulfed, or would she drive before it? Happily, she forged ahead, though shaking like a spent race-horse, and with every timber straining and creaking. Now, thrown broadside on, her side was remorselessly battered by the floe pieces; then, tossed by the sea over ice-block after ice-block, she seemed like a plaything in the grasp of an irresistible power. For some hours this severe trial of strength and fortitude endured; then the storm subsided as rapidly as it had arisen, and their gratitude for their own escape was mingled with joy at the safety of the *Dorothea*, which they could see in the distance, still afloat, and with her crew in safety.

On Captain Parry's second expedition, in 1822, his ships, the *Hecla* and the *Fury*, were placed in a position of scarcely less danger.

Thus we read of the *Hecla*, which at the time had been made fast by means of cables to the land-ice, that a very heavy and extensive floe caught her on her broadside, and, being backed by another large body of ice, gradually lifted her stern as if by the action of a wedge. The weight every moment increasing, her crew were obliged to veer on the hawsers, whose friction was so great as nearly to cut through the bitt-heads, and ultimately set them on fire, so that it became requisite to pour upon them buckets of water. At length the pressure proved irresistible; the cables snapped; but as the sea was too full of ice to allow the ship to drive, the only way in which she could yield to the enormous burden brought to bear upon her was by leaning over the land-ice, while her stern at the same time was lifted clean out of the water for fully five feet.

Had another floe backed the one which lifted her, the ship must inevitably have rolled broadside over, or been rent in twain. But the pressure which had been so dangerous eventually proved its safety; for, owing to its increasing weight, the floe on which she was carried burst upwards, unable to resist its force. The *Hecla* then righted, and a small channel opening up amid the driving ice, she was soon got into comparatively smooth water.

On the following day, shortly before noon, a heavy floe, measuring some miles in length, came down towards the *Fury*, exciting the gravest apprehensions for her safety. In a few minutes it came in contact, at the rate of a mile and a half an hour, with a point of the land-ice, breaking it up with a tremendous roar, and forcing numberless immense masses, perhaps many tons in weight, to the height of fifty or sixty feet; whence they again rolled down on the inner or land side, and were quickly succeeded by a fresh supply. While they were compelled to remain passive spectators of this grand but terrific sight, being within five or six hundred yards of the point, the danger they incurred was twofold: first, lest the floe should swing in and serve the ship in the same unceremonious manner; and, secondly, lest its pressure should detach the land-ice to which they were secured, and cast them adrift at the mercy of the tides. Fortunately, neither of these terrible alternatives occurred, the floe remaining stationary for the rest of the tide, and setting off with the ebb when the tide soon afterwards turned.

The reader must not imagine that an ice-field is a smooth and uniform plain, as level as an English meadow; it is, on the contrary, a rugged succession of hollows, and of protuberances called "hummocks," interspersed with pools of water, and occasionally intersected by deep fissures. In many parts it can be compared only to a promiscuous accumulation of rocks closely packed together, and piled up over the extensive dreary space in great heaps and endless ridges, leaving scarcely a foot of level surface, and compelling the traveller to thread his way as best he can among the perplexing inequalities; sometimes mounting unavoidable obstructions to an elevation of ten, and again more than a hundred feet, above the general level.

The interspaces between these closely accumulated ice-masses are filled up to some extent with drifted snow.

Now, let the reader endeavour to form a definite idea of the scene presented by an ice-field. Let him watch the slow progress of the sledges as they wind through the labyrinth of broken ice-tables, the men and dogs pulling and pushing up their respective loads, as Napoleon's soldiers may have done when drawing their artillery through the rugged Alpine passes, or Lord Napier's

AMONG THE ICE-HUMMOCKS.

heroes when they scaled the steep Abyssinian heights. He will see them clambering over the very summit of lofty ridges, where no gap occurs, and again descending on the other side, the sledge frequently toppling over a precipice, sometimes capsizing, and sometimes breaking.

Again: he will see the adventurous party, when baffled in their attempt to cross or find a pass, breaking a track with shovel and handspike; or, again, unable even with these appliances to accomplish their end, they retreat to seek an easier route. Perhaps they are fortunate enough to discover a kind of gap or gateway, and upon its winding and uneven surface accomplish a mile or so with comparative ease. The snow-drifts sometimes prove an assistance, but more frequently an obstruction; for though their surface is always hard, it is not always firm to the foot. Then the crust gives way, and the foot sinks at the very moment when the other is lifted. But, worse than this, the chasms between the hummocks may be overarched with snow in such a manner as to leave a considerable space at the bottom void and empty; then, when everything looks auspicious, down sinks one of the hapless explorers to his waist, another to the neck, a third is "lost to sight," the sledge gives way, and all is confusion worse confounded! To educe order out of the chaos is probably the work of hours; especially if the sledge, as is often the case, must be unloaded. Not unfrequently it is necessary to carry the cargo in two or three loads; the sledges are coming and going continually; and the day is one "endless pull and haul."

Dr. Hayes speaks of an ice-floe, crested with hummocks, and covered with crusted snow, the solid contents of which he estimated, in round numbers, at 6,000,000,000 of tons, its depth being about one hundred and sixty feet. All around its border was banked up a kind of rampart of last year's ice, the loftiest pinnacle of which rose fully one hundred and twenty feet above the sea-level. This ice-tower consisted of blocks of ice of every shape and size, piled one upon another in the greatest disorder. Numerous other towers, or bastions, equally rugged, though of less elevation, sprang from the same ridge, and from every part of this desolate area; and "if a thousand Lisbons were crowded together and tumbled to pieces by the shock of an earthquake, the scene could hardly be more rugged, nor to cross the ruins a severer task."

We must date the origin of a floe like this back to a very remote period. Probably it was cradled, at the outset, in some deep recess of the land, where it remained until it had accumulated to a thickness which defied the summer's sun and the winter's winds. Then it would grow, as the glacier grows, from above; for, like the glacier, it is wholly composed of fresh ice—that is, of frozen snow. Thus it will be seen, to quote Dr. Hayes once more, that the accumulation of ice upon the mountain-tops is in nowise different from the accumulation which takes place upon these floating fields, where every recurring year marks an addition to their depth. Vast as they are to the sight, and pigmies as they are compared with the inland Mer de Glace, yet, in all that concerns their growth, they are truly glaciers, dwarf floating glaciers. That only in this manner can they grow to so great a depth will at once be conceded by the reader, if he recollects that ice soon reaches a maximum thickness by direct freezing, and that its growth is arrested by a natural law. Necessarily, this maximum thickness varies according to the temperature of the locality: but the ice is in itself the sea's protection. The cold air cannot absorb the warmth of the water through more than a certain thickness of ice, and that thickness attains a final limit long before the winter has reached its close. The depth of ice formed on the first night is greater than that formed on the second; on the second is greater than on the third; on the third greater than on the fourth; and so it continues, until the increase no longer takes place. In other words, the ratio of increase of the thickness of ice is in inverse proportion to the duration of the period of freezing. There comes a time when the water beneath the ice no longer congeals, because the ice-crust above it protects it from the action of the atmosphere. Dr. Hayes asserts that he never saw an Arctic ice-table *formed by direct freezing* that exceeded eighteen feet; and he justly adds, that were it not for this all-wise provision of the Deity,—this natural law, as our men of science term it,—the Arctic waters would, ages ago, have been solid seas of ice to their profoundest depths.

Having said thus much about the various forms which the ice assumes in the Polar seas;—about their icebergs and ice-fields, pack-ice and drift-ice, and the thick belt of ice which surrounds their shores,—we may now direct the reader's attention to their Animal Life; to the creatures which inhabit them, walrus and seal and whale, the fishes, the molluscs, and even minuter organisms.

And first we shall begin with the Walrus, which finds a congenial home in the Arctic wildernesses.

Walrus-hunting is the principal, or at all events the most lucrative, occupation of the Norse fishermen, who annually betake themselves to the cheerless shores of Spitzbergen in search of booty. Their life is a terribly hard and dangerous one; and Mr. Lamont, who has had much experience of them, observes that they all have a restless, weary look about the eyes,—a look as if contracted by being perpetually in the presence of peril. They are wild, rough, and reckless; but they are also bold, hardy, and enduring of cold, hunger, fatigue; active and energetic while at sea, though sadly intemperate during their winter-holiday.

The vessels engaged in the seal-fishery and walrus-hunting are fitted out by the merchants of Tromsöe and Hammerfest, who have, of late years, adopted the system of sharing their proceeds with their crews, thus giving them a direct interest in the prosperity of the expedition. The ship is fitted out and provisioned by the owners, who also advance to the men what money they may require to purchase clothing and to make provision for their families during their absence. Then they allot one-third of the gross receipts of the adventure to the crew, dividing it into shares, three for the captain, two for the harpooneer, and one each for the common men. So that if a fairly successful voyage should realize in skins, blubber, and ivory a sum of two thousand dollars, and the number of hands amounts to ten, the usual strength of a seal-ship's crew, each will receive forty-seven and a half dollars, or about £10,—a very considerable sum for a Norwegian.

Each ship carries a couple of boats, and a walrus-boat, capable of holding five men, which measures twenty-one feet in length by five feet beam, having her main breadth at about seven feet from the bow. She is bow-shaped at both ends, and so built as to turn easily on her own centre, besides being strong, light, and easy to row. Each man plies a pair of oars hung in "grummets" to stout thole-pins; the steersman directs the boat by also rowing a pair of oars, but with his face to the bow; and as there are six thwarts, he can, if necessary, sit and row like the others. By this arrangement the strength of the men is economized, and the boat is more swiftly turned when in pursuit of the walrus.

The steersman also acts as harpooneer, and, of course, sits in the bow. The strongest man in the boat is usually placed next to him, to hold and haul in the line when a walrus is struck, and it is his duty to hand the harpoons and lances to the harpooneer as required.

Each boat—which, by the way, is painted white, so as to resemble the ice amongst which it moves—is usually provided with three harpoon-heads inside the bow, on each side: these fit into little racks of painted canvas, so that their keen points and edges may not be blunted, and to prevent them from injuring the men. The harpoons serve equally well for seal and walrus, and, simple as they seem and are, answer admirably the purpose for which they are designed. The weapon is thrust into the animal; its struggles tighten the line; the large outer barb then catches up a loop of its tenacious hide, or the tough reticulated fibres containing its blubber;

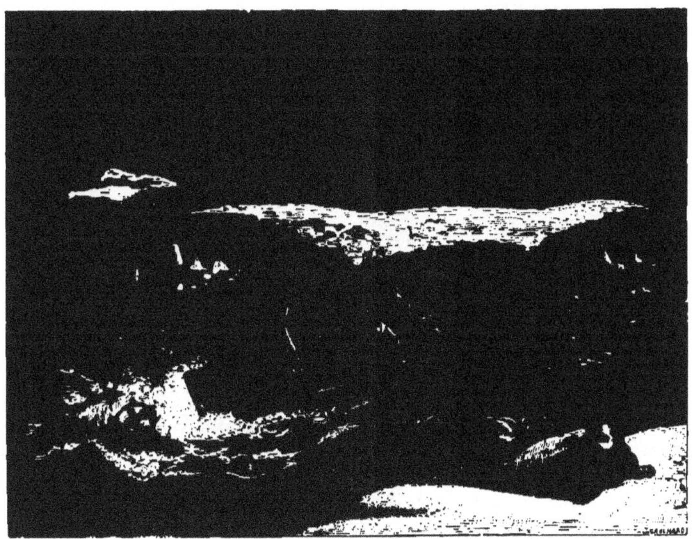

HUNTING THE WALRUS.

while the small inner barb, like that of a fish-hook, prevents it from being detached or loosened. When a walrus has been properly struck, and the line hauled taut, it rarely escapes. To each harpoon a line of twelve or fifteen fathoms long is attached: a sufficient length, as the walrus is seldom found in water more than fifteen fathoms deep; and even if the water should exceed that depth, it cannot drag the boat under, because it is unable to exert its full strength when subjected to the pressure of twelve or fifteen fathoms of water.

Besides the harpoons, each boat is provided with four or five enormous lances; the shaft being made of pine-wood, nine feet long, and one inch and a half thick at the handle, increasing upwards to a thickness of two inches and a half where it enters the iron socket. This would seem a formidable weapon, and formidable it is in the stout hands of a Norse harpooneer; yet, frequently, the iron shank is bent double, or the strong shaft snapped like a reed, in the violent resistance of the sea-horse; and, therefore, to prevent the head being lost, it is fastened to the shaft by a double thong of raw seal-skin, tied round the shank and nailed to the handle for about three feet up. The shaft may seem of disproportionate length, but it is necessary to give the buoyancy sufficient for floating the heavy iron spear if it should fall into the water. This spear, or lance, is not used for seals, because it would spoil the skins.

Notwithstanding the destruction effected by the yearly expeditions of the walrus-hunters, the sea-horses are still found in large herds in many parts of the Polar world. Mr. Lamont describes a curious and exciting spectacle, where four large flat icebergs were seen to be so closely packed with these animals that they were sunk almost level with the water, and presented the appearance of "solid islands of walrus!" The walrus lay with their heads reclining on one another's backs and hind-quarters, just as rhinoceroses lie asleep in the dense shade of the African forests, or, to use a more commonplace but familiar comparison, as hogs slumber and wallow in a British farmyard.

Such a sight was a temptation not to be withstood by a walrus-hunter, and Mr. Lamont and his harpooneer speedily disturbed the repose of the monsters, which chiefly consisted of cows and young bulls. After slaying their victims, and getting them on board, came the disagreeable but necessary task of separating the blubber from the skins to stow it in the barrels; a process which is performed in the following manner:—

Across the ship's deck, immediately aft the hatchway, is erected a kind of framework or stage of stout timber, about four feet in height, but sloping down at an angle of about sixty. degrees, with the deck at the forward side: on the other side it is perpendicular, and there the two *specksioneers* (or "blubber-cutters") post themselves, clad, not in armour, but in oil-skin from top to toe, and armed with large keen knives, curved on the edge. Then the skins are hoisted out of the hold, and, two at a time, are suspended across the frame, with the blubber side uppermost: the fat, or blubber, is next removed by a kind of *mowing* motion of the knife, which is held in both hands, and swayed from left to right. Only long practice, and great steadiness of wrist, can give the dexterity requisite for the due performance of this difficult operation. Even in skinning a walrus, skill is imperative.

As the blubber is mown off, it is divided into slabs, weighing twenty or thirty pounds each, and flung down the hatchway, where two men are stationed to receive it, and pack it into the casks, which when full are securely fastened up.

The skin, which is taken off the animal in two longitudinal halves, is a valuable commodity, and sells at the rate of from two to four dollars per half skin. The principal purchasers are the Russian and Swedish merchants, and its principal uses are for harness and sole leather. It is also twisted into tiller ropes, and employed to protect the rigging of ships from friction. The blubber is valued on account of the oil; but neither has the walrus so much blubber, in proportion to its size, as the seal, nor does the blubber afford so good an oil. A seal of 600 lbs. will carry 200 to 250 lbs. weight of fat; an ordinary walrus, weighing 2000 lbs., will not carry any more.

The most profitable portion of the unfortunate sea-horse is its tusks, which are composed of very hard, dense, and white ivory. This ivory is not so good, and consequently does not command so high a price, as elephant ivory, but is in high repute for the manufacture of false teeth, chessmen, umbrella handles, whistles, and other small articles.

The tusks are not an extra pair of teeth, but a development and modification of the canines. For about six or seven inches of their length they are solidly set in the mass of hard bone which forms the animal's upper jaw. So far as they are imbedded in the head they are hollow, but mostly filled up with a cellular osseous substance containing much oil; the remainder of the tusk is hard and solid throughout.

The young walrus, or calf, has no tusks in its first year of existence; but in its second, when it is about the size of a large seal, it has a pair of much the same size as the canines of a lion. In the third year the tusks measure about six inches in length.

In size and shape they vary greatly, according to the animal's age and sex. A good pair of bull's tusks, says Mr. Lamont, will be twenty-four inches each in length, and four pounds each in weight; but larger and heavier specimens are of frequent occurrence. Cows' tusks, it is said, will average fully as long as those of the bulls, because less liable to be broken, but seldom weigh more than three pounds. They are generally set much closer together than the bull's tusks, sometimes even overlapping one another at the points; while those of the bull will often diverge as much as fifteen inches.

In scientific language the walrus, morse, or sea-horse (*Trichecus*), belongs to a genus of amphibious mammals of the family *Phocidæ*, a family including the well-known seals. It agrees with the other members of that family in the general configuration of the body and limbs, but distinctly differs from them in the head, which is remarkable,—as we have seen,—for the extraordinary development of the canine teeth of the upper jaw, as also for the protuberant or swollen appearance of the muzzle,—due to the size of their sockets and the thickness of the upper lip. This upper lip is thickly set with strong, transparent, bristly hairs, which measure about six inches in length, and are as thick as a crow-quill. The terrific moustache, with the long white curving tusks, the thick projecting muzzle, and the fierce and bloodshot eyes, give *Rosmarus trichecus* a weird and almost demoniacal aspect as it rears its head above the waves, and goes far to account for some of the legends of sea-monsters which embellish the Scandinavian mythology.

THE WALRUS, OR MORSE.

The walrus has no canine teeth in the lower jaw. Its incisors are small, and ten in number; six in the upper and four in the lower jaw. The molars, at first five on each side in each jaw, but fewer in the adult, are simple and not large; their crowns are obliquely worn. The nostrils would seem to be displaced by the sockets of the tusks; at least they both open almost directly upwards at some distance from the muzzle. The eyes are small, but savage; there are no external ears.

The Arctic walrus is the sole known species of the genus. It is a gregarious animal, always assembling in large herds, which occasionally leave the water to take their rest upon the shore or

A WALRUS FAMILY.

on the ice; and it is at such times the hunters chiefly attack them, since their movements *out* of the water are very laborious and awkward.

They defend themselves against their enemies, of which the Polar bear is chief, with their formidable tusks; and these they also use in their fierce combats with one another. They fight with great determination and ferocity, using their tusks much in the same manner as game-cocks use their beaks. From the unwieldy appearance of the animal, and the position of its tusks, an inexperienced spectator would suppose that the latter could be employed only in a *downward* stroke; but, on the contrary, it turns its neck with so much ease and rapidity that it can strike in all directions with equal force.

FIGHT BETWEEN A WALRUS AND A POLAR BEAR.

Old bulls very fre-

BOAT ATTACKED BY WALRUS.

quently have one or both of their tusks broken; which may arise either from fighting or from using them to assist in scaling the rocks and ice-floes. But these broken tusks are soon worn down again and sharpened to a point by the action of the sand, as the walrus, like the elephant, employs its tusks in digging its food out of the ground,—that is, out of the ocean-bed. Its food principally consists of starfish, shrimps, sandworms, clams, cockles, and algæ; and Scoresby relates that he has found the remains of young seals in its stomach.

In reference to the gradual decay, or, more correctly speaking, extermination of the walrus, the following particulars seem to be authentic.

When the pursuit of the walrus was first systematically organized from Tromsöe and Hammerfest, much larger vessels were employed than are now in vogue; and it was usual for them to obtain their first cargo about Bear Island early in the season, and two additional cargoes at Spitzbergen before the summer passed away. This regular and wholesale slaughter drove away the sea-horse herds from their haunts about Bear Island; but even afterwards it was not a rare occurrence to procure three cargoes in a season at Spitzbergen, and less than two full cargoes was regarded as a lamentable mishap. Now, however, more than one cargo in a season is very seldom obtained, and many vessels return, after four months' absence, only half full.

It is estimated that about one thousand walrus and twice that number of bearded seals (*Phoca barbata*) are annually captured in the seas about Spitzbergen, exclusive of those which sink or may die of their wounds. Some idea, therefore, may be formed of the number of sea-horses which still ride the waves of the Polar seas. But it is quite clear that they are undergoing a rapid diminution of numbers, and also that they are gradually withdrawing into the inaccessible solitudes of the remotest North.

We learn from the voyage of Ohthere, which was undertaken ten centuries ago, that the walrus then abounded even on the very coast of Finmarken. They have abandoned that region, however, for some centuries, though individual stragglers were captured up to within the last forty years. After their desertion of Finmarken, they retreated to Bear Island; thence they were driven to the Thousand Islands, Hope Island, and Ryk-Yse Island; and thence, again, to the banks and skerries to the north of Spitzbergen. It is fortunate for the persecuted walrus that the latter districts are accessible only in open seasons, or perhaps once in every three or four summers; so that they obtain a respite and time to breed and replenish their numbers. Otherwise the end of the present century would mark also the total extinction of the walrus on the island-shores of Northern Europe.

We agree with Dr. Kane that the resemblance of the walrus to man has been absurdly overstated. Yet the notion is put forward in some of our systematic treatises, and accompanied by the suggestion that we are to look for the type of the merman and mermaid in this animal. If we look we shall not find. The walrus has a square-shaped head, with a frontal bone presenting a steep descent to the eyes, and any likeness to humanity must exist in the imagination of the spectator. Some of the seals exhibit a much greater resemblance: the size of the head, the regularity of the facial oval, the drooping shoulders, even the movements of the seal, remind us impressively of man. And certainly, when seen at a distance, with head raised above the waves, it affords some justification for the fanciful conception of the nymphs of ocean, the mermaids who figure so attractively in song and legend.

Dr. Kane remarks that the instinct of attack, which is strong in the walrus, though so feeble in the seal, and is a well-known characteristic of the pachyderms, is interesting to the naturalist, as assisting to establish the affinity of the walrus to the latter. When wounded, it rears its body high out of the water, plunges heavily against the ice, and strives to raise itself upon the surface by means of its fore-flippers. As the ice gives way under its weight, its countenance assumes a truly ferocious expression, its bark changes to a roar, and the foam pours out from its jaws till it froths its beard.

Even when not excited, the walrus manages its tusks bravely. So strong are they that they serve as grappling-irons with which to hold on to the surface of the steep rocks and ice-banks it loves to climb; and thus it can ascend rocky islands that are sixty and a hundred feet above the sea-level. It can deal an opponent a fearful blow, but it prefers to charge, like a veteran warrior; and man, unless well armed, often comes off second best in the contest.

FIGHT WITH A WALRUS.

Governor Flaischer told Dr. Kane that, in 1830, a brown walrus—and the Eskimos say that the brown walrus are the fiercest—after being speared and wounded near Upernavik, put to flight its numerous assailants, and drove them in fear to seek help from the Danish settlement. So violent were its movements as to jerk out the harpoons that were launched into its body. The governor slew it with much difficulty after it had received several rifle-shots and lance-wounds from his whale-boat.

On another occasion, a young and adventurous Innuit plunged his *nalegeit* into a brown walrus; but, alarmed by the savage demeanour of the beast, called for help before using the lance. In vain the older and more wary hunters advised him to forbear. "It is a brown walrus!" they cried; "*Aúvok-Kaiok!* Hold back!" Finding the caution disregarded, his only brother rowed forward, and hurled the second harpoon. Almost instantaneously the infuriated beast charged, like the wild boar, on the unfortunate young Innuit, and ripped open his body.

Here is a description of a walrus-hunt:—

On first setting out, the hunters listen eagerly for some sounds by which to discover the habitat of the animal. The walrus, like amateur vocalists, is partial to its own music, and will lie for hours enjoying the monotonous vocalization in which it is accustomed to indulge. This is described as something between the mooing of a cow and the deepest baying of a mastiff; very round and full, with its "barks" or "detached notes" repeated seven to nine times in rather quick succession.

The hunters hear the bellow, and press forward in single file; winding behind ice-hummocks and ridges in a serpentine approach towards a group of "pond-like discolorations," recently frozen ice-spots, which are surrounded by older and firmer ice.

In a few minutes they come in sight of the walrus. There they are, five in number, rising at intervals through the ice in a body, and breaking it up with an explosion which sounds like the report of heavy ordnance. Conspicuous as the leaders of the herd are two large and fierce-looking males.

Now for a display of dexterity and skill. While the walrus remains above water, the hunter lies flat and motionless; when it begins to sink, behold, the hunter is alert and ready to spring. In fact, scarcely is the tusked head below the water-line before every man is in a rapid run; while, as if by instinct, before it returns all are prone behind protecting knolls of ice. They seem to guess intuitively, not only how long it will be absent, but the very point at which it will reappear. And, in this way, hiding and advancing by turns, they reach a plate of thin ice, scarcely strong enough to bear a man's weight, on the very brink of the dark pool in which the walrus are gambolling.

The phlegmatic Eskimo harpooneer now wakens into a novel condition of excitement. His coil of walrus-hide, a well-trimmed line of many fathoms length, lies at his side. He attaches one end to an iron barb, and this he fastens loosely, by a socket, to a shaft of unicorn's horn; the other end is already loosed. It is the work of a second! He has grasped the harpoon. The water eddies and whirls; puffing and panting, up comes the unwieldy sea-horse. The Eskimo rises slowly; his right arm thrown back, his left hanging close to his side. The walrus looks about him, and throws the water off his crest; the Eskimo launches the fatal weapon, and it sinks deep into the animal's side.

Down goes the wounded *awuk*, but the Eskimo is already speeding with winged feet from the scene of combat, letting his coil run out freely, but clutching the final loop with a desperate grip. As he runs, he seizes a small stick of bone, roughly pointed with iron, and by a swift strong movement thrusts it into the ice; he twists his line around it, and prepares for a struggle.

The wounded walrus plunges desperately, and churns the ice-pool into foam; meantime, the line is hauled tight at one moment, and loosened the next; for the hunter has kept his station. But the ice crashes; and a couple of walrus rear up through it, not many yards from the spot where he stands. One of them, a male, is excited, angry, partly alarmed; the other, a female, looks calm, but bent on revenge. Down, after a rapid survey of the field, they go again into the ocean-depths; and immediately the harpooneer has chosen his position, carrying with him his coil, and fixing it anew.

Scarcely is the manœuvre accomplished before the pair have once more risen, breaking up an area of ten feet in diameter about the very spot he left. They sink for a second time, and a second time he changes his place. And thus continues the battle between the strength of the beast and the address of the man, till the former, half exhausted, receives a second wound, and gives up the contest.

The Eskimos regard the walrus with a certain degree of superstitious reverence, and it is their belief that it is under the guardianship of a special representative or prototype, who does not, indeed, interfere to protect it from being hunted, but is careful that it shall be hunted under tolerably fair conditions. They assert that near a remarkable conical peak, which rises in the solitudes of Force Bay, a great walrus lives all alone, and when the moon is absent, creeps out to the brink of a ravine, where he bellows with a voice of tremendous power.

The walrus-hunter, unless he keeps to the sea-shore, and the ice-floes within reach of a boat, must be prepared to undergo many hardships, and to confront with a calm heart the most baffling and terrible dangers. He may be overtaken by a gale; and a gale in the wild remote North, far from any shelter,—a gale which drives before it the blinding snow and pitiless icicles,—a gale which sweeps unresisted and irresistible over leagues of frozen snow,—a gale which comes down from the mountain-recesses where the glaciers take their rise,—is something so dread, so ghastly, that the dweller in temperate regions can form no idea of it.

We remember that one of the gallant seekers after Franklin describes an Arctic gale, and its effects. He says that the ice, at a short distance from the shore, had in many places been swept bare of snow by the driving blast; and over the glassy sheet he and his companions were helplessly carried along before the gale. The dogs, seldom stretching their traces, ran howling in front of the sledges, which pressed upon their heels.

Wild was the scene, and dark. The moon had sunk far behind the snow-shrouded mountains, and the travellers had no other light than the shimmer of stars. The deep shadows of the cliffs, towering a thousand feet above their heads, lay heavily upon them, and enhanced the midnight gloom. The patches of snow clinging to the sharp angles of the colossal wall; the white shroud lying on its lofty summit; the glaciers which here and there protruded through its clefts, brought out into striking relief the blackness of its cavernous recesses. The air was filled with clouds of drift, which sometimes completely hid the land, and swept relentlessly before the explorers, as they tottered across the frozen plain.

Suddenly a dark line became visible across their path; its true nature revealed by circling wreaths of "frost-smoke." "*Emerk! emerk!*" (Water! water!) shouted the drivers, checking as suddenly as possible the headway of the sledges, but not until the party were within a few feet of a recently opened and rapidly widening crack,—a fissure in the ice-crust, already twenty feet across.

Some of the travellers now clambered to the summit of a pile of hummocks, and endeavoured to pierce the obscurity. A headland, laid down on the map as Cape Alexander, lay only a few miles in advance. The ice in the shallow bay on its southern side was rent in all directions; while beyond, from the foot of the cape, a broad sheet of water extended westward. The wind diversified its dark surface with ridges of snowy spray; while here and there a frosty surf tumbled in breakers over a small berg or drifting floe. The pieces of ice lying along its margin were in motion, and the crash of their hard surfaces could be heard as they came into constant collision. Their strident clamour, the ceaseless washing of the surface, the moaning of the wind, the steely rush of the drift, the piteous wail of the dogs, and all the strange noises and voices of the storm, added to the gloom and awful melancholy of that moonless night.

We need not wonder that the Eskimos of the Arctic wilderness are as fearful of a tempest as are the Bedouins of the African desert. It overwhelms the one with a cloud of snow, and it buries the other in a cloud of sand; and each demands and receives its quota of victims.

That seal-hunting should be more extensively pursued than walrus-hunting is natural; for if less exciting, it is also less dangerous; and the seal is not only a more valuable prey than the walrus, but is more easily captured.

The Phocidæ are well represented in the Arctic waters. In Behring Sea we encounter the sea-lion and the sea-bear; while from the Parry Islands to Novaia Zemlaia extends the range of the harp seal (*Phoca Grœnlandica*), the bearded seal (*Phoca barbata*), and the hispid seal (*Phoca hispida*). The skins of all these species are more or less valuable; their oil is much esteemed; and their flesh supplies the wild northern tribes with one of their principal articles of subsistence.

The structure of the seal is admirably adapted in every detail to an aquatic life. It lives

HERD OF SEALS, NEAR THE DEVIL'S THUMB, BAFFIN SEA, GREENLAND.

chiefly in the water, where its motions are always easy and graceful; but it spends a part of its time in enjoying the sunshine on ice-fields, open shores, rocks, and sandy beaches; and the female brings forth her young on land.

The body of the seal is elongated, and tapers considerably from the chest to the tail. The head has been compared to that of the dog; the brain is generally voluminous. The feet are short, and little more than the paw extends beyond the integument of the body; they are webbed, and pentadactylous, or five-toed: the fore feet are set like those of other quadrupeds; but the hind feet are directed backwards, with toes which can be spread out widely to act as paddles. The tail is short.

The motions of the seal on land are constrained and peculiar. The fore feet are but little used, and the body is thrown forward in a succession of jerks produced by a contraction of the spine. Awkward as this mode of progression seems, it is, nevertheless, exceedingly rapid. The seal, however, never ventures far from the shore, and the moment it is disturbed or alarmed it plunges into the water.

The physiognomy of the animal is in perfect accord with its character, and expresses a considerable degree of intelligence combined with much mildness of disposition. The eyes are large, black, and brilliant; the nose is broad, with oblong nostrils; and there are large whiskers. The seal has no external ears, but in the auricular orifices exists a valve which can be closed at will, and protects the internal organism from the water; the nostrils possess a similar valve. The body is thickly garnished with stiff glossy hairs, very closely set against the skin, and plentifully lubricated with an oily secretion, so that the surface is always smooth, and unaffected by water. The teeth differ in different genera, but in all are specially adapted for the seizure of fish and other slippery prey, though the seals are omnivorous in their habits, and will partake both of vegetable and animal food. There are either six or four incisors in the upper, and four or two in the lower jaw; the canines are invariably large and strong; and the molars, usually five or six on either side, in each jaw, are sharp-edged or conical, and bristle with points. The seal is fond of swallowing large stones; for what purpose is not certain, but, probably, to assist digestion.

Seals live in herds, more or less numerous, along the frozen shores of the Arctic seas; and on the lonely deserted coasts they bring forth their young, over which they watch with singular affection. They swim with much rapidity, and can remain a considerable time under water. They are migratory in their habits, and at least four species visit our British waters. On the northern coasts of Greenland they are observed to take their departure in July and to return again in September. They produce two or three young at a time, and suckle them for six or seven weeks in remote caverns and sequestered recesses; after which they take to the sea. The young exhibit a remarkable degree of tractability; will recognize and obey the maternal summons; and assist each other in distress or danger. Many, if not all, of the species are polygamous, and the males frequently contend with desperate courage for the possession of a favourite female.

There is not much difference in the habits of the different genera or species of the Phocidæ; but while the great Arctic seal dives like the walrus, making a kind of semi-revolution as it goes down, the common seal (*Phoca vitulina*), called by the hunters the *stein-cobbe*, from its custom of basking on the rocks, dives by suddenly dropping under water, its nose being the last part of its body which disappears, instead of its tail.

The common seal has a very fine spotted skin, and weighs about sixty or seventy pounds. It is much fatter, in proportion to its size, than the bearded seal, and its carcass, consequently, having less specific gravity, floats much longer on the water after death.

A third kind of seal found in the Spitzbergen seas is, probably, the Phoca hispida, though the hunters know it only by the names of the "springer," and Jan Mayen seal. In the spring months it is killed in large numbers by the whalers among the vast ice-fields which encircle the solitary rocks of Jan Mayen Island.

Mr. Lamont observes that these seals, though existing in such enormous numbers to the west, are not nearly so numerous in Spitzbergen as the great, or even as the much less abundant common seal. They are gregarious, which neither of the other varieties are, and generally consort in bands of fifty to five hundred. They are extremely difficult to kill, as during the summer months they very seldom go upon the ice; they seem much less curious than the other seals, and go at such a rapid pace through the water as to defy pursuit from a boat. On coming up to breathe, these seals do not, like their congeners, take a deliberate breath and a leisurely survey, but the whole troop make a sort of simultaneous flying leap through the air like a shoal of porpoises, as they go along, and reappear again at an incredible distance from their preceding breathing-place. Hence the name of "springers" given to them by the whalers.

THE COMMON SEAL.

The Jan Mayen seal weighs from 200 to 300 lbs., and is described as the fattest and most buoyant of the Arctic mammals.

We have spoken of seal's flesh as an important article of subsistence to the Eskimo tribes. Our Arctic voyagers and explorers have frequently been glad to nourish themselves upon it, and speak of it as somewhat resembling veal in flavour. Not once or twice, but several times, it has saved the hardy pioneer of civilization from destruction, and the discovery of a stray seal has been the means of preserving a whole expedition.

There is a very striking incident of this kind in the narrative of Dr. Kane. He and his party had reached Cape York on their way to the Danish settlements, after their long but fruitless search for Sir John Franklin. They were spent with fatigue, and half-dead from hunger. A kind of low fever crippled their energies, and they were unable to sleep. In their frail and unseaworthy boats, which were scarcely kept afloat by constant bailing, they made but slow progress across the open bay; when, at this crisis of their fortunes, they descried a large seal floating, as is the wont of these animals, on a small patch of ice, and apparently asleep,—a seal so large that at first they mistook it for a walrus.

Trembling with anxiety, Kane and his companions prepared to creep down upon the monster.

One of the men, Petersen, with a large English rifle, was stationed in the bow of the boat, and stockings were drawn over the oars as mufflers. As they approached the animal, their excitement became so intense that the men could hardly keep stroke. That no sound might be heard, Dr. Kane communicated his orders by signal; and when about three hundred yards off the oars were taken in, and they moved on, stealthily and silently, with a single scull astern.

The seal was not asleep, for he reared his head when his enemies were almost within rifle-shot; and long afterwards Dr. Kane could remember the hard, careworn, almost despairing expression of the men's haggard faces as they saw him move; their lives depended on his capture. Dr. Kane lowered his hand, as a signal for Petersen to fire. M'Gorry, who was rowing, hung, he says, upon his oar, and the boat slowly but noiselessly forging ahead, did not seem within range. Looking at Petersen, he saw that the poor fellow was paralyzed by his anxiety, and was vainly seeking to find a rest for his gun against the cut-water of the boat. The seal rose on his flippers, gazed at his antagonists for a moment with mingled curiosity and alarm, and coiled himself for a plunge. At that moment, simultaneously with the crack of the rifle, he relaxed his huge bulk on the ice, and, at the very brink of the water, his head fell helplessly on one side.

SHOOTING A SEAL.

Dr. Kane would have ordered another shot, but no discipline could have controlled his men. With a wild yell, each vociferating according to his own impulse, they urged both boats upon the floes. A crowd of hands seized the precious booty, and bore it up to safer ice. The men seemed half crazy, they had been so reduced by famine. They ran over the floe, crying and laughing, and brandishing their knives. Before five minutes had elapsed, each man was sucking his streaming fingers or mouthing long strips of raw blubber.

Not an ounce of this seal was wasted!

The intestines found their way into the soup-kettles without any observance of the preliminary home-processes. The cartilaginous parts of the fore-flippers were cut off in the *mêlée*, and passed round for the operation of chewing; and even the liver, warm and raw as it was, bade fair to be eaten before it had seen the pot. That night, on the large halting-floe to which, in contempt of the dangers of drifting, the happy adventurers had hauled their boats, two entire planks of the *Red Eric* were devoted to the kindling of a large cooking-fire, and they enjoyed a bountiful and savage feast.

Such is an experience of Arctic life; of the hardships endured by the heroic men who go forth to do the work of Science and Civilization.

Returning to the seals, we may remark that, according to a scientific authority, the angle of woody rock on which a phoca is accustomed to rest with his family comes to be regarded as his property, and no other individuals of his species are entitled to lay claim to it. Although in the water these animals congregate together in numerous herds, and protect and courageously defend one another, yet, when they have once emerged from their favourite element, they regard themselves on their own space of rock as in a sacred domicile, where no comrade has a right to intrude on their domestic tranquillity. If any stranger approach this family centre, the chief—or shall we call him the father?—prepares to repel by force what he considers an unwarrantable encroachment; and a terrible combat invariably ensues, which terminates only with the death of the lord of the rock, or the compulsory retreat of the intruder.

But a family never seizes upon a larger tract than it absolutely requires, and lives peaceably with neighbouring families, from which it is seldom separated by a greater interval than forty or fifty paces. If compelled by necessity, they will even live on amicable terms at much closer quarters. Three or four families will share a rock, a cavern, or an ice-floe; but each occupies the place allotted to it at the original apportionment, and shuts himself within it, so to speak, nor ever meddles with individuals of another family.

THE OTARY.

Our modern naturalists divide the Phocidæ into two distinct orders: the *Phocæ* properly so called, which have no external ears, but only an auditory orifice on the surface of the head; and the *Otariæ*, which are provided with external organs.

The remarks we have been making apply more particularly to the common seal (*Phoca vitulina*), or small Spitzbergen seal, which measures from four to five feet in length. The Greenland or harp seal (*Phoca Grœnlandica*), to which we have already alluded, is larger and fatter, and is distinguished by the changes of colour it undergoes before it reaches maturity. We have also spoken of the bearded seal (*Phoca barbata*), which sometimes attains a length of ten feet, and is known, not only by its size, but its thick and strong moustaches. The hooded seal (*Stemmatopus cristatus*) is distinguished by the globular and expansible sac situated on the summit of the head of the males. This species grows to the length of seven or eight feet, and inhabits the waters of Newfoundland and Greenland.

The value of the seal to the Eskimo tribes will best be understood from a description of the uses to which various parts of the animal are applied in an Eskimo hut.

We will suppose this hut to measure about five or five and a half feet in height, and about ten feet in diameter. The walls are made of stones, moss, and the bones of seals, narwhals, whales, and other ocean-creatures. They are not arched, but recede inward gradually from the foundation, and are capped by long oblong slabs of slate-stone extending from side to side. We enter: the flooring consists of thin flat stones. At the back part of the hut the floor rises about a foot, and this *breck*, as the elevation is called, serves both as couch and seat, being covered with a thick layer of dried moss and grass, under seal-skins, dog-skins, and bear-skins. Similar elevations are placed at the corners in front; under one of which will lie, perhaps, a litter of pups, with their mother, and under the other a portion of seal's meat. In the square front of the hut, above the passage-way, a window is inserted; the light being admitted through a square sheet of strips of dried intestines, sewed together. The entrance is in the floor, close to the front wall, and is covered with a piece of seal-skin. Seal-skins are hung about the walls to dry. At the edge of the *breck*, on either side, sits a woman, each busily engaged in attending to a smoky lamp, fed with seal's oil. These lamps are made of soapstone, and in shape resemble a clam-shell, being about eight inches in diameter. The cavity is filled with oil obtained from seal's blubber; and on the straight edge the flame burns quite vividly, the wick which furnishes it being made of moss. The business of the women is apparently to prevent the lamps from smoking, and to keep them supplied with blubber, large pieces of which are placed in the cavity, the heat drawing out the oil. About three inches above this flame hangs, suspended from the ceiling, an oblong square pot made of the same material as the lamp, in which a joint of seal is simmering slowly. Above this hangs a rack, made of bare rib-bones, bound together crosswise, on which stockings and mittens, and various garments made of seal-skin, are laid to dry. No other fire can be seen than that which the lamps supply, nor is any other needed. So many persons are crowded into the confined interior that it is insufferably hot, while the whole place reeks with the smell of seal-flesh, seal-oil, and seal-skin!

THE HOODED SEAL.

It is natural enough that we should here introduce an account of the Eskimo mode of catching seals. The great season of the seal-hunt is the spring, when the inoffensive *phocæ* gambol and sport in the open water-ways near the coasts, or clamber on the ice-floes to enjoy the rays of the tardy sun. They are of a wary and timid disposition, and we may suppose that their traditions have taught them to be on their guard against man; but as all their habits and ways are well known to the Eskimo, they do not succeed in eluding his dexterous perseverance. Sometimes the hunter attires himself in a seal-skin, and so exactly imitates their appearance and movements that he approaches within spear-range of them before the disguise is detected; or else he creeps into their haunts behind a white screen, which is propelled in front of him by means of a sledge. As the season verges upon midsummer less precaution becomes necessary; the eyes of the seals being so congested by the fierce radiance of the sun that they are often nearly blind. In winter they are assailed while labouring at their breathing-holes, or when they rise for the purpose of respiration.

If an Eskimo satisfies himself that a seal is working away beneath the ice, he takes up his station at the suspected point, and seldom quits it, however severe the weather, until he has captured the animal. To protect himself from the freezing blast, he throws up a snow-wall about

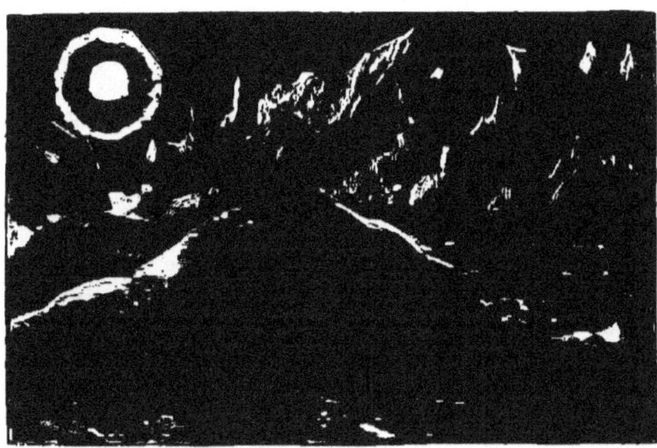

AN ESKIMO SEAL-HUNTER.

four feet in height, and seating himself in its shade, he rests his spears, lines, and other appliances on a number of little forked sticks inserted into the snow, in order that he may move them, when wanted, without making the slightest noise. He carries his caution to such an excess, that he even ties his own knees together with a thong to prevent his garments from rustling!

To discover whether the seal is still gnawing at the ice, our patient watcher makes use of his *keep-kuttuk*; a slender rod of bone, no thicker than ordinary bell-wire, cleverly rounded, with a knob at one end and a sharp point at the other.

This implement he thrusts into the ice, and the knob, which remains above the surface, informs him by its motion whether the animal is still engaged in making his hole; if it does not move, the attempt is given up in that place, and the hunter betakes himself elsewhere. When he supposes the hole to be nearly completed, he stealthily raises his spear, and as soon as he can hear the blowing of the seal, and knows therefore that the ice-crust is very thin, he drives it into the unsuspecting animal with all his might; and then hacks away with his sharp-edged knife, or *panna*, the intervening ice, so as to repeat his blows, and secure his victim. The *neituk*, or Phoca hispida, being the smallest seal, is held while struggling, either by the hand, or by a line one end of which is twisted round a spear driven into the ice. In the case of the bearded seal, or *oguka*, the line is coiled round the hunter's leg or arm; for a walrus, round his body, the feet being at the same time firmly planted against a hummock of ice, so as to increase the capability of resistance. A boy of fifteen can kill a *neituk*, but the larger animals can be mastered only by a robust and experienced adult.

We come now to speak of the Whale, which, in size, is the sovereign of the Arctic seas, and the grandest type of marine life.

Whales (*Cetacea*) are, as most persons now-a-days know, an order of aquatic mammals, distinguished by their fin-like anterior extremities, and by the peculiarity that the place of the posterior extremities is supplied by a large horizontal caudal fin, or tail; while the cervical bones are so compressed that the animal, externally at least, seems to have no neck.

The general form of the whale, notwithstanding its position among the Mammalia, is similar to that of the fishes, and the horizontal elongation of the body, the smooth and rounded surface, the gradual attenuation of the extremities of the trunk, and the magnitude of the fins and tail, are specially adapted to easy and swift motion in the water. The arrangement of the bones composing the anterior limb is very curious. The whole of the fin consists of exactly the same parts as those which we find in the human hand and arm; but they are so concealed beneath the thick cutaneous or integumentary envelope, that not a trace of bone is visible. In this respect an intermediate organization is shown by the fore limbs of the seal.

The posterior extremity, in all the Cetacea, is either absolutely deficient, or else rudimentary. If rudimentary, its sole vestige consists of certain small bones, the imperfect representation of a pelvis, suspended, as it were, in the flesh, and unconnected with the spinal column. Here we may observe a remarkable difference between the whale and the seal: in the latter, as we have seen, there is a short tail, and the posterior extremities perform the office of a true caudal fin; in the former this important organ of progression consists, to use Mr. Bell's words, of "an extremely broad and powerful horizontal disc, varying in figure in the different genera, but in all constituting the principal instrument of locomotion." In fishes the tail is set vertically, but in whales horizontally; and it has been well said that the admirable adaptation of such a peculiarity in its position to the requirements of the animal forms a fresh and beautiful illustration of the infinite resource and foresight of the Creative Wisdom.

Thus: the fishes, respiring only the air contained in the dense liquid medium in which they live, require no access to the atmosphere; and, therefore, their progression is chiefly confined to the same region. But the whales, breathing atmospheric air, must necessarily come to the surface for each respiration; and hence they need a powerful instrument or lever, the position of which shall apply its impulse in a vertical direction, so as to impel their colossal bulk from the lowest depths of ocean to the surface every time the lungs require to receive a fresh supply of atmospheric air. The greatest rapidity of motion is effected by alternate strokes of the tail against the water, upwards and downwards; but the usual progression is accomplished by an oblique lateral and downward impulse, first on one side and then on the other, just as a boat is propelled by a man with a single oar in the art of "sculling." The extent of the tail in some of the larger species is really immense; the superficies being no less than about a hundred square feet, and its breadth considerably exceeding twenty feet.

The common, right, or Greenland whale (*Balæna mysticetus*) has been, for centuries, the object of man's systematic pursuit, on account of its valuable oil and scarcely less valuable baleen.

This whale seldom exceeds fifty to sixty feet in length, or thirty to forty in girth, and, therefore, is by no means the head of its family. As in other species, the body is thick and bulky forwards, largest about the middle, and tapers suddenly towards the tail. The head is colossal; broad, flat, and rounded beneath, and narrow above; it forms about a third of the animal's entire length, and is about ten or twelve feet broad. Its lips—such lips!—are five or six feet thick. They do not cover any teeth, but they protect a pair of very formidable jaws. The cavernous

THE GREENLAND WHALE.

interior of the mouth is filled up with two series of whalebone laminæ, about three hundred in each, which require particular description. The whalebone, or baleen, as it is called, consists of numerous parallel plates, layers, or laminæ, each of which is formed of a central coarse fibrous layer lying between two that are compact and externally polished. But this outer part does not completely cover the inner; a kind of edge is exposed, and this edge terminates in a loose fringed or fibrous extremity. Moreover, at the base of each plate of baleen lies a conical cavity, covering a pulp which corresponds with it; and this pulp is sunk within the substance of the gum or buccal membrane stretched over the palate and upper jaw.

The compact *outer* layers of the baleen plate are continuous with a white horny layer of the gum, which passes on to the surface of each plate; and the pulp may be regarded, therefore, as the secreting organ of the internal coarse structure only. The filaments of the fringe are exceedingly numerous, and so fill up the mouth-cavity as to form a very efficient and ingenious sieve or strainer; and as the esophagus, or "swallow," of the whale is so confined as to be unable to admit of the passage even of the smaller fish, and the food of the whale consequently is limited to minute organisms, such as the medusæ, this skilfully devised construction is absolutely requisite in order to retain the whole of those which are taken into the mouth.

The mode in which the whale feeds may be thus described :—

The broad waters of the Arctic seas teem with innumerable shoals of molluscous, radiate, and crustaceous animals, and these are frequently so numerous as absolutely to colour the wave-surface.

When a whale, therefore, desires food, it opens its colossal mouth, and a host of these organisms is, as it were, swept up by the great expanse of the lower jaw: as the mouth closes, the water is ejected, and the life it contained is imprisoned by the appliance we have attempted to describe.

If we consider the number of whales found in the Northern seas, and the mighty bulk of each individual, our imagination entirely fails to appreciate the countless myriads of minute organisms which must be sacrificed to their due nourishment.

One of the principal products of the Greenland whale is its baleen, or whalebone, with the domestic uses of which our readers will be familiar; but the large quantities of oil which it yields are still more valuable. A whale sixty feet in length will supply fully twenty tons of pure oil.

Besides the common whale, our hunters find in the seas of the North the razor-backed whale, or northern rorqual (*Balænoptera physalis*), characterized by the prominent ridge which extends along its mighty back. This monster of the deep attains a length, it is said, of one hundred feet, and measures from thirty to thirty-five feet in circumference. But its yield of oil and baleen is less than that of the right or Greenland whale, and as its capture is a task of difficulty and danger, the whalers seldom attack it. In its movements it is more rapid and restless, and when harpooned it frequently plunges downward with such force and velocity as to break the line. In several respects it differs from the Greenland species; and particularly in the nature of its food, for it feeds upon fishes of considerable size.

Some of our naturalists affirm that several species of rorquals exist in the Arctic seas; and the pike whale, so called from the resemblance of its mouth to that of a pike, is frequently described as an independent species. Others, however, are of opinion that the pike is simply the young of the monster we have been describing. The rorqual is very voracious, and preys extensively upon fishes; as many as six hundred cod, to say nothing of smaller "fry," having been found in the stomach of a single individual.

While the Greenland whale is being rapidly driven back into the icy wildernesses beyond Behring Strait, on the west, and the creeks and gulfs beyond Baffin Bay, on the east, the rorquals, including the *Balænoptera rostratus* (or beaked whale), *Balænoptera musculus*, and *Balænoptera boöps*, still frequent the open waters,—their pursuit being, as we have shown, more difficult and less profitable. They are generally found in attendance on the herring-shoals, of which they are the assiduous and destructive enemies. Off Greenland, Spitzbergen, and Novaia Zemlaia they are found in considerable numbers.

Our whalers go forth every year in well-provided ships, and supplied with the best and most formidable weapons which scientific ingenuity can devise. Still they find the enterprise one of peril and hardship, and it is universally recognized as requiring in those who embark in it no ordinary powers of endurance, as well as courage, patience, and perseverance. Yet the Asiatic and American tribes do not fear to confront the ocean-leviathan with the simplest of arms. The Aleüt embarks in his little skiff, or *baidar*, and catching sight of his prey, stealthily approaches it from behind until he nearly reaches the monster's head. Then he suddenly and dexterously drives his short spear into the huge flank, just under the fore fin, and retreats as swiftly as his well-plied oars can carry him. If the spear has sunk into the flesh, the whale is doomed; within the next two or three days it will perish, and the currents and the waves will hurl the vast bulk on the nearest shore, to be claimed by its gallant conqueror. And as each spear bears its owner's peculiar mark, the claim is never disputed.

Occasionally the baidar does not escape in time, and the exasperated leviathan, furiously lashing the waters with its tail, hurls the frail boat high up into the air, as if it were a reed, or sinks it with one crushing blow. No wonder that those of their race who undertake so hazardous a calling are held in high repute among the Aleüts. To sally forth alone, and encounter the whale in the icy waters of the Polar Sea, is a task demanding the utmost intrepidity and the utmost tranquillity of nerve.

Many of the whales thus daringly harpooned are lost. It is on record that, in the summer of 1831, one hundred and eighteen whales were struck near Kadjack, and of these only forty-three were found. The others either drifted to far-off shores and lonely unknown isles, or became the prey of sharks and ocean-birds. Wrangell states that of late years the Russians have introduced the use of the harpoon, and engaged some English harpooneers to teach the Aleüts the secret of their craft; and, therefore, the older and more hazardous method, which the Aleüts had learned from their forefathers, will soon be a thing of the past.

The Eskimos devote the month of August to the whale-fishery, and for this purpose they assemble in companies, and plant a colony of huts on some bold headland of the Polar coast, where the water is of depth sufficient to float their destined victim.

As soon as a whale's colossal bulk is seen outstretched on the water, a dozen kayaks or more cautiously paddle up in the rear, until one of them, shooting ahead, comes near enough on one side for the men to drive the spear into its flesh with all the force of both arms. To the spear are attached an inflated seal-skin and a long coil of thong. The whale dives immediately it is stricken. After awhile it reappears, and the signal being given by the floating seal-skin buoy, all the canoes again paddle towards their prey. Again the opportunity is seized for launching the fatal spears; and this process is repeated until the exhausted whale rises more and more frequently to the surface, is finally killed, and towed ashore.

Captain M'Clure fell in with an Eskimo tribe off Cape Bathurst which hunted the whale in this primitive fashion, but the females, as well as the men, engaged in the pursuit. An *omaiak*, or woman's boat, he says, is "*manned* by ladies," having as harpooneer a chosen man of the tribe; and a shoal of small fry, in the form of *kayaks*, or single-men canoes, are in attendance. The harpooneer singles out "a fish," drives into its flesh his weapon, to which an inflated seal-skin is attached by means of a walrus-hide thong. The wounded fish is then incessantly harassed by the men in the kayacks with weapons of a similar description; and a number of these, driven into the unfortunate whale, baffle its efforts to escape, and wear out its strength, until, in the course of a day, it dies from exhaustion and loss of blood.

Sherard Osborn tells us that the harpooneer, when successful, becomes a very great personage indeed, and is invariably decorated with the Eskimo order of the Blue Ribbon; that is, a blue line is drawn across his face over the bridge of his nose. This is the highest honour known to the heroes of Cape Bathurst; but it carries along with it the privilege of the decorated individual being allowed to take unto himself a second wife!

In the waters of Novaia Zemlaia, Greenland, and Spitzbergen is found the narwhal, or sea unicorn (*Monodon monoceros*), which was at one time the theme of so many extravagant legends. It belongs to the Cetacea, but differs from the whale in having no teeth, properly so called, and in being armed with a formidable horn, projecting straight forward from the upper jaw, in a direct line with the body. This horn, or tusk, the use of which has not been satisfactorily ascertained, is harder and whiter than ivory, spirally striated from base to point, tapers throughout, and measures from six to ten feet in length. Mr. Bell remarks that it would be a strange anomaly

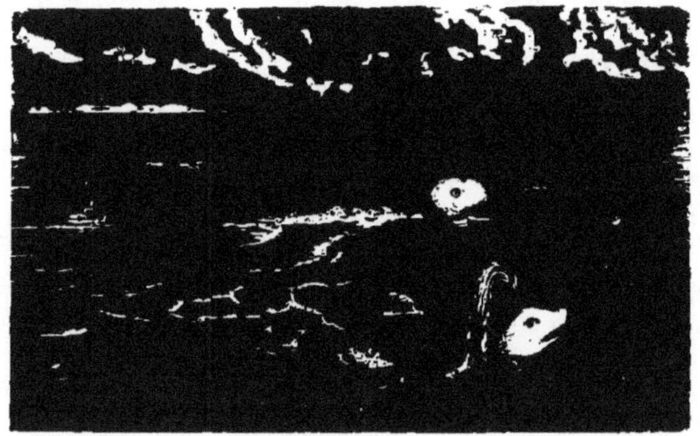

NARWHALS, MALE AND FEMALE.

if the apparent singleness of this weapon were real. In truth, *both* teeth are invariably found in the jaw, not only of the male, but of the female also; but in ordinary (though not in all) cases one only, and this in the male, is fully developed, the other remaining in a rudimentary condition —even as both do in the female.

The narwhal, from mouth to tail, is about twenty feet long, though individuals measuring thirty feet are sometimes met with. Its head is short, and the upper part convex; its mouth small; its spiracle, or respiratory vent, duplicate within; its tongue long; the pectoral fins small. The back, which is convex and rather wide, has no fins, and sharpens gradually towards the tail, which, as in other Cetacea, is horizontal. The food of the narwhal, whose habits are remarkably pacific, consists of medusæ, the smaller kinds of flat fish, and other marine animals.

A striking spectacle which frequently greets the eye of the voyager in the Arctic seas is that of a shoal of dolphins gambolling and leaping, as if in the very heyday of enjoyment. The beluga, sometimes called the white whale (*Delphinus leucos*), attracts attention by the dazzling whiteness of its body and the swiftness of its movements. It frequents the estuaries of the Obi and the Irtish, the Mackenzie and the Coppermine, which it sometimes ascends to a considerable distance in pursuit of the salmon. Its length varies from twelve to twenty feet; it has no dorsal fin; and its head is round, with a broad truncated snout.

The black dolphin (*Globicephalus globiceps*) is also an inhabitant of the Polar seas, both beyond Behring Strait, and between Greenland and Spitzbergen. It is, however, frequently met with in waters further south. Its length averages about twenty-four feet, and its circumfer-

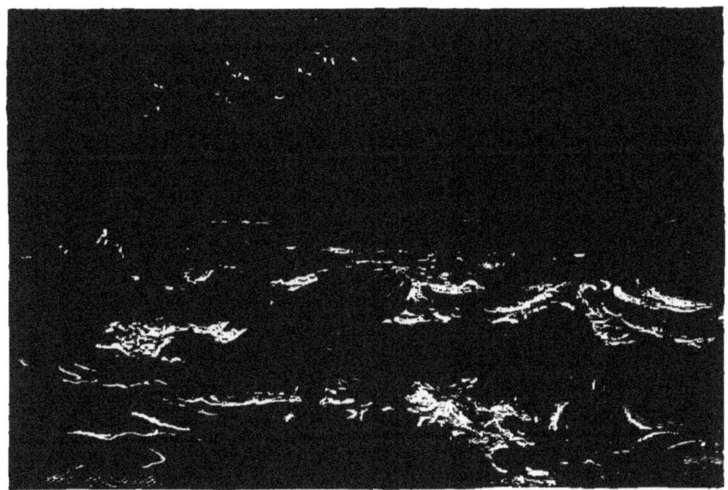

A SHOAL OF DOLPHINS.

ence ten feet. Its smooth oily skin is bluish-black on the upper, and an obscure white on the lower, parts of the body. Twenty-two or twenty-four strong interlocking teeth in each jaw form its formidable apparatus of offence and defence; its dorsal fin is about fifteen inches high; its tail five feet broad; the pectoral fins are long and narrow, and well adapted to assist their owner in its rapid movements. It consorts with its kind in herds of several hundreds, under the guidance of some old and wary males, whom the rest follow as docilely as a flock of sheep their bell-wether; hence the Shetlanders term it the "ca'ing whale." Large shoals are frequently stranded on the shores of Norway, Iceland, and the Orkney, Faroe, and Shetland Isles, furnishing the inhabitants with a welcome booty.

To the same latitudes belong the ferocious orc or grampus (*Delphinus orca*), the tiger of the seas, which not only attacks the porpoise and dolphin, but even the colossal whale. Its broad deep body is black above and white beneath; the sides are marbled with black and white. There are thirty teeth in each jaw, those in front being blunt, round, and slender, while those behind are sharp and thick; and between each is a space fitted to receive those of the opposite jaw when the mouth is closed. The back fin of the grampus is of great size; sometimes measuring as much as six feet in length, from the base to the tip. The grampus generally voyages in small squadrons of four or five individuals, following each other in single file, and alternately rising and sinking in such a manner as to resemble the undulatory motions of a huge kraken or sea-serpent.

Among the inhabitants of the Polar Ocean must certainly be included the Polar bear (*Thalassarctos maritimus*), since it swims and dives with great dexterity, and, moreover, is often found on the drifting ice-floes at a distance of eighty to one hundred miles from land. It is a creature of great strength, great fierceness, and great courage, though we may not accept the exaggerated accounts of it which enliven the narratives of the earlier voyagers.

A noble creature is the Polar bear, says Sherard Osborn, whether we speak of him by the

POLAR BEARS.

learned titles of "Ursus maritimus," "Thalassarctos maritimus," or the sailors' more expressive nomenclature of "Jack Rough!" With all her many wonders, continues this lively writer, never did Nature create a creature more admirably adapted to the life it has to lead. Half flesh, half fish, the seaman wandering in the inhospitable regions of the North cannot but be struck with the appearance of latent energy and power its every action attests, as it rolls in a lithe and swaggering way over the rough surface of the frozen sea; or, during the brief Arctic summer, haunts the broken and treacherous "pack" in search of its prey.

When not too loaded with fat—and it seems to fatten readily—the pace of the bear is leisurely and easy, yet at its slowest it is equal to that of a good pedestrian; and when alarmed or irritated, its speed is surprising, though not graceful. On level ice, it flings itself ahead, as it were, by a violent jerking motion of the powerful fore paws, in what has been described as an "ungainly gallop;" but it always makes, when it can, for rough ice, where its strength and agility are best displayed, and where neither man nor dog can overtake it. In the Queen's Channel, during Captain M'Clure's expedition, more than one bear was seen making its way over broken-up ice, rugged and precipitous as the mind can picture, with a truly wonderful facility; their powerful fore paws and hind legs enabling them to spring from piece to piece, scaling one fragment and sliding down another with the activity of a huge quadrumane rather than that of a quadruped. Evidently it is conscious of its superiority in such rough and perilous ground, and is generally found at the edge of the belts of hummocks or broken ice which intersect most ice-fields, or else amongst the frozen pack-ice of channels such as Barrow's and the Queen's.

There is, however, another reason why bears keep among hummocks and pack-ice—namely, that near such spots water usually first makes its appearance in the summer. Seals, consequently, are most numerous there; while the inequalities of the floe afford shelter to the bears in approaching their prey. During summer the colour of the Polar bear is of a dull yellowish hue, closely resembling that of decaying snow or ice. The fur is then thin, and the hair on the soles of their feet almost wholly rubbed off, as with the other animals of Arctic climes; but in the autumn, when the body has recovered from the privations of the previous winter, and a thick coating of blubber overlays his carcass to meet the exigencies of another season of scanty fare, the feet, as the season advances, are beautifully incased and feathered with hair, and the animal's colour usually turns to a very pale straw, which, from particular points of view, as the light strikes it, looks white, or nearly so. The nose and lips are of a jetty black; the eyes vary in colour. Brown is common, but some have been seen with eyes of a pale gray. Their sense of smell is peculiarly acute, facilitated no doubt by the peculiar manner in which the pure keen air of the North carries scent to very considerable distances.

Sherard Osborn states that bears have been seen to follow up a scent, exactly as dogs would do; and the floes about Lowther Island, in 1851, looked as if the bears had quartered there in search of seals, after the fashion of a pointer in the green fields of England. The snorting noise which they make as they approach near indicates how much more confidence they place in their scent than in their vision; though both, when the hunter is concerned, are apt to deceive them.

The Polar bear attains to very formidable proportions; but when seamen speak of monsters fifteen feet in length, their auditors may be excused for withholding their belief. *Ten* feet would seem to be a maximum; and the bear need be large, strong, and muscular to master the large Arctic seal, especially the saddle-back and bladder-nose species. For though it swims well and dives well, it neither swims nor dives as well as the seal, and would therefore have but little chance of obtaining a sufficient livelihood if it could not attack and capture its victim on the ice-floes.

The seal, on the other hand, fully aware of its danger, and of the only means of escaping from it, always keeps close to the water, whether it be the hole it has gnawed and broken through the ice, or the open sea at the floe edge.

And when it lies basking on the floating ice, and apparently apathetic and lethargic, nothing can exceed its vigilance. With its magnificent eyes it is able to sweep a wide range of the horizon, however slightly it turns its head; its keenness of hearing adds to its security. There is something peculiarly striking in its continuous watchfulness. Now it raises its head and looks around; now it is intent on the slightest sound that travels over the crisp surface of the ice; now it gazes and listens down its hole, a needful precaution against so subtle a hunter as old Bruin!

It would seem impossible to surprise an animal so vigilant and so wary; and, indeed, in circumventing its prey the bear exhibits an astuteness and a skill which overpass the bounds of instinct, and approach closely to those of reason.

From its scent and by its quick strong vision the bear apprehends the position of the seal. Then it throws itself prone upon the ice, and profiting by inequalities which are invisible to human eyes, gradually steals upon its destined victim by a soft and scarcely perceptible movement of the hind feet. To hide its black muzzle, it constantly uses its fore feet; and thus, only the dingy white of its coat being visible, it is scarcely to be distinguished from the general mass of the floe. Patiently it draws nearer and nearer; the seal, mistaking it for one of its own congeners, or else yielding to a fatal curiosity, delaying until its assailant, with one spring, is upon it.

BEAR CATCHING A SEAL.

Yet, as the old adage says, there is many a slip; and even in these circumstances the bear does not always secure its feast. It is disappointed sometimes just as the prey seems within its grasp; and how keen the disappointment is can be appreciated only, we are told, by hapless Arctic travellers, "who have been hours crawling up, dreaming of delicious seal's fry and overflowing fuel bags, and seen the prey pop down a hole when within a hundred yards of it." The great muscular power of the seal frequently enables it to fling itself into the water in spite of the bear's efforts to hold it on the floe; Bruin, however, retains his grip, for his diving powers are not much inferior to those of the seal, and down they go together! Sometimes the bear proves victorious, owing to mortal injuries inflicted upon the seal before it reaches the water; sometimes it may be seen reappearing at another hole in the floe, or clambering up another loose piece of ice, apparently much mortified by its want of success.

As we have said, the bear dives well, and is nearly as much at home in the water as upon the ice. If it catches sight of a seal upon a drifting floe, it will slide quietly into the sea, swim with only the tip of its nose above the water, and, diving under the floe, reach the very spot which the hapless seal has regarded as an oasis of safety. It is this stratagem of its enemy which has taught the seal to watch its hole so warily. Even on extensive ice-fields fast to the land, where the bear cannot conceal its approach by taking advantage of hummocks or other inequalities, the seal is not safe; for then Bruin drops down a hole, and swims along under the ice-crust until it reaches the one where the poor seal is all unwittingly enjoying its last rays of sunshine.

The bear's season of plenty begins with the coming of the spring. In February and March the seal is giving birth to her young, who are born blind and helpless, and for ten days are unable to take to the water. The poor mothers use every effort to protect them, but, in spite of their affectionate exertions, a perfect massacre of the innocents takes place, in which, not improbably, the Arctic wolf is not less guilty than the Arctic bear.

Voracity, however, frequently proves its own Nemesis, and the bear, in its eager pursuit of prey, often involves itself in serious disaster. The seal instinctively breeds as close as possible to the open water. But the ice-floes, during the early equinoctial gales, will sometimes break up and drift away in the form of pack-ice; a matter of indifference, says Osborn, to the seal, but a question of life and death to the bear. Borne afar on their little islets of ice, rocked by tempestuous waters, buffeted by icy gales, numbers of these castaways are lost along the whole area of the Polar Sea. It is said that when the gales blow down from the north, bears are sometimes stranded in such numbers on the shores of Iceland as to endanger the safety of the flocks and herds of the Icelandic peasants; and they have been known to reach the coasts of Norway.

Bears drifting about at a considerable distance from the land are often enough seen by the whalers. They have been discovered fully sixty miles from shore, in Davis Strait, without any ice in sight, and utterly exhausted by long swimming. It is thus that Nature checks their too rapid increase; for beyond the possibility of the wolf hunting it in packs and destroying the cubs, there seems no other limitation of their numbers. The Eskimos are too few, and too badly provided with weapons, to slaughter them very extensively. Wherever seals abound, so do bears; in Barrow Strait and in the Queen's Channel they have been seen in very numerous troops. The Danes assert that they are plentiful about the northern settlement of Upernavik in Greenland, for nine months in the year; and from the united testimony of the natives inhabiting the north-eastern portion of Baffin Bay, and that of Dr. Kane, who wintered in Smith Sound, it is evident that they are plentiful about the *polynias*, or open pools, formed there by the action of the tides.

In the summer months, when the bear is loaded with fat, it is easily hunted down, for then it can neither move swiftly nor run long; but in deep winter its voracity and its great strength render it a formidable enemy to uncivilized and unarmed man. Usually it avoids coming into contact with our British seamen, though instances are on record of fiercely contested engagements, in which Bruin has with difficulty been defeated.

It is folly, says Sherard Osborn, to talk of the Polar bear hibernating: whatever bears may do on the American continent, there is only one Arctic navigator who ever saw a bear's nest! Bears were seen at all points visited by our sailors in the course of M'Clure's expedition; at all times and in all temperatures; males or females, and sometimes females with their cubs. In mid-winter, as well as in mid-summer, they evidently frequented spots where tides or currents occasioned either water to constantly exist, or only allowed such a thin coating of ice to form that the seal or walrus could easily break through.

That the Polar bear does not willingly attack man, except when hotly pursued or when suffering from extreme want, is asserted by several good authorities, and confirmed by an experience which Dr. Hayes relates. He was strolling one day along the shore, and observing with much interest the effect of the recent spring-tides upon the ice-foot, when, rounding a point of land, he suddenly found himself confronted in the full moonlight by an enormous bear. It had just sprung down from the land-ice, and met Dr. Hayes at full trot, so that they caught sight of each other, man and brute, at the same moment. Being without a rifle or other means of defence, Dr. Hayes suddenly wheeled towards his ship, with much the same reflections, probably, about discretion and valour as occurred to old Jack Falstaff when the Douglas set upon him; but discovering, after a few lengthy strides, that he was not "gobbled up," he looked back over his shoulder, when, to his gratification as well as surprise, he saw the bear speeding towards the open water with a celerity which left no doubt as to the state of its mind. It would be difficult to determine which, on this occasion, was the more frightened, the bear or Dr. Hayes!

A curious illustration of the combined voracity and epicureanism of Bruin is recorded by Dr. Kane. A *cache*, or depôt of provisions, which had been constructed by one of his exploring parties with great care, and was intended to supply them with stores on their return journey, they found completely destroyed. It had been built, with every possible precaution, of rocks brought together by heavy labour, and adjusted in the most skilful manner. So far as the means of the builders permitted, the entire construction was most effective and resisting. Yet these "tigers of the ice" seemed to have scarcely encountered an obstacle. Not a morsel of pemmican (preserved meat) remained, except in the iron cases, which, being round, with conical ends, defied both claw and teeth. These they had rolled and pawed in every direction, — tossing them about like footballs, although upwards of eighty pounds in weight. An alcohol - case, strongly iron-bound, was dashed into small fragments; and a tin can of liquor twisted almost into a ball. The bears' strong claws had perforated the metal, and torn it up as with a chisel.

BEARS DESTROYING A CACHE.

But the burglars were too dainty for salt meats. For ground coffee they had evidently a relish; old canvas was also a favourite,—*de gustibus non est disputandum*; even the flag which had been reared "to take possession" of the icy wilderness, was gnawed down to the very staff. It seemed that the bears had enjoyed a regular frolic; rolling the bread-barrels over the ice-foot and into the broken outside ice; and finding themselves unable to masticate the heavy India-rubber cloth, they had amused themselves by tying it up in unimaginable hard knots.

The she-bear displays a strong affection for her young, which she will not desert even in the extremity of peril. The explorer already quoted furnishes an interesting narrative of a pursuit of mother and cub, in which the former's maternal qualities were touchingly exhibited.

On the appearance of the hunting party and their dogs, the bear fled; but the little one being unable either to keep ahead of the dogs or to maintain the same rate of speed as its mother, the latter turned back, and, putting her head under its haunches, threw it some distance forward. The cub being thus safe for the moment, she would wheel round and face the dogs, so as to give it a chance to run away; but it always stopped where it had alighted, until its mother

FIGHT WITH A WHITE BEAR.

came up, and gave it another forward impulse; it seemed to expect her aid, and would not go forward without it. Sometimes the mother would run a few yards in advance, as if to coax her cub up to her, and when the dogs approached she would turn fiercely upon them, and drive them back. Then, as they dodged her blows, she would rejoin the cub, and push it on,—sometimes putting her head under it, sometimes seizing it in her mouth by the nape of its neck.

For some time she conducted her retreat with equal skill and celerity, leaving the two hunters far in the rear. They had sighted her on the land-ice; but she led the dogs in-shore, up a small stony valley which penetrated into the interior. After going a mile and a half, however, her pace slackened, and, the little one being spent, she soon came to a halt, evidently determined not to desert it.

At this moment the men were only half a mile behind; and, running at full speed, they soon reached the spot where the dogs were holding her at bay. The fight then grew desperate. The mother never moved more than two yards ahead, constantly and affectionately looking at her cub. When the dogs drew near, she sat upon her haunches, and taking the little one between her hind legs, she fought her assailants with her paws, roaring so loudly that she could have been heard a mile off. She would stretch her neck and snap desperately at the nearest dog with her shining teeth, whirling her paws like the sails of a windmill. If she missed her aim, not daring to pursue one dog lest the others should pounce upon her cub, she uttered a deep howl of baffled rage, and on she went, pawing and snapping, and facing the ring, grinning at them with wide-opened jaws.

When the hunters came up, the little one apparently had recovered its strength a little, for it was able to turn round with its dam, however quickly she moved, so as always to keep in front of her belly. Meantime the dogs were actively jumping about the she-bear, tormenting her like so many gadflies; indeed, it was difficult to fire at her without running the risk of killing the dogs. But Hans, one of the hunters, resting on his elbow, took a quiet, steady aim, and shot her through the head. She dropped at once, and rolled over dead, without moving a muscle.

Immediately the dogs sprang towards her; but the cub jumped upon her body and reared up, for the first time growling hoarsely. They seemed quite afraid of the little creature, she fought so actively, and made so much noise; and, while tearing mouthfuls of hair from the dead mother, they would spring aside the minute the cub turned towards them. The men drove the dogs off for a time, but were compelled to shoot the cub at last, as she would not quit the body.

A still more stirring episode is recorded by Dr. Kane, which will fitly conclude our account of the Polar bear.

"*Nannook! nannook!*" (A bear! a bear!) With this welcome shout, Hans and Morton, two of his attendants, roused Dr. Kane one fine Saturday morning.

To the scandal of his domestic regulations, the guns were all impracticable. While the men were loading and capping anew, Dr. Kane seized his pillow-companion six-shooter, and ran on deck, to discover a medium-sized bear, with a four-months' cub, in active warfare with the dogs. They were hanging on her skirts, and she, with remarkable alertness, was picking out one victim after another, snatching him by the nape of the neck, and flinging him many feet, or rather yards, by a scarcely perceptible movement of her head.

Tudea, the best dog, was already *hors de combat*; he had been tossed twice. Jenny, another of the pack, made an extraordinary somerset of nearly fifty feet, and alighted senseless. Old Whitey, a veteran combatant, stanch, but not "bear-wise," had been foremost in the battle; soon he lay yelping, helplessly, on the snow.

It seemed as if the battle were at an end; and *nannook* certainly thought so, for she turned aside to the beef-barrels, and began with the utmost composure to turn them over, and nose out their fatness. A bear more innocent of fear does not figure in the old, old stories of Barents and the Spitzbergen explorers.

Dr. Kane now lodged a pistol-ball in the side of the cub. At once the mother placed her little one between her hind legs, and, shoving it along, made her way to the rear of the store or "beef-house." As she went she received a rifle-shot, but scarcely seemed to notice it. By the unaided efforts of her fore arms she tore down the barrels of frozen beef which made the triple walls of the store-house, mounted the rubbish, and snatching up a half barrel of herrings, carried it down in her teeth, and prepared to slip away. It was obviously time to arrest her movements. Going up within half pistol-range, Dr. Kane gave her six buck-shot. She dropped, but instantly rose, and getting her cub into its former position, away she sped!

And this time she would undoubtedly have effected her escape, but for the admirable tactics of Dr. Kane's canine Eskimo allies. The Smith Sound dogs, he says, are educated more thoroughly than any of their more southern brethren. Next to the seal and the walrus, the bear supplies the staple diet of the tribes of the North, and, except the fox, furnishes the most important element of their wardrobe. Unlike the dogs Dr. Kane had brought with him from Baffin Bay, the Smith Sound dogs were trained, not to attack, but to embarrass. They revolved in circles round the perplexed bear, and when pursued would keep ahead with regulated gait, their comrades accomplishing a diversion at the critical moment by a nip at the *nannook's* hind-quarters. This was done in the most systematic manner possible, and with a truly wonderful composure. "I have seen bear-dogs elsewhere," says Dr. Kane, "that had been drilled to relieve each other in the *mêlée*, and avoid the direct assault; but here, two dogs, without even a demonstration of attack, would put themselves before the path of the animal, and retreating right and left, lead him into a profitless pursuit that checked his advance completely."

The unfortunate animal was still fighting, and still retreating, embarrassed by the dogs, yet affectionately carrying along her wounded cub, and though wounded, bleeding, and fatigued, gaining ground upon her pursuers, when Hans and Dr. Kane secured the victory, such as it was, for their own side, by delivering a couple of rifle-balls. She staggered in front of her young one, confronted her assailants in death-like defiance, and did not sink until pierced by six more bullets.

When her body was skinned, no fewer than nine balls were discovered. She proved to be of medium size, very lean, and without a particle of food in her stomach. Hunger, probably, had stimulated her courage to desperation. The net weight of the cleansed carcass was 300 pounds; that of the entire animal, 650 pounds; her length, only 7 feet 8 inches.

It is said that bears in this lean condition are more palatable and wholesome than when fat; and that the impregnation of fatty oil through the cellular tissues makes a well-fed bear nearly uneatable. The flesh of a famished beast, though less nutritious as body-fuel or as a stimulating diet, is rather sweet and tender than otherwise. *Moral:* starve your bear before you eat him!

The little cub was larger than the qualifying adjective would imply. She was taller than a dog, and her weight 114 lbs. She sprang upon the corpse of her slaughtered mother, and rent the air with woful lamentations. All efforts to noose her she repelled with singular ferocity; but at last, being completely muzzled with a line fastened by a running knot between her jaws and the back of her head, she was dragged off to the brig amid the uproar of the dogs.

Dr. Kane asserts that during this fight, and the compulsory somersets which it involved, not a dog suffered seriously. He expected, from his knowledge of the hugging propensity of the plantigrades, that the animal would rear, or if she did not rear, would at least use her fore arms; but she invariably seized the dogs with her teeth, and after disposing of them for a time, refrained from following up her advantage,—probably because she had her cub to take care of. The Eskimos state that this is the habit of the hunted bear. One of the Smith Sound dogs made no exertion whatever when he was seized, but allowed himself to be flung, with all his muscles relaxed, a really fearful distance; the next instant he rose and renewed the attack. According to the Eskimos, the dogs soon learn this "possum-playing" habit.

It would seem that the higher the latitude, the more ferocious the bear, or that he increases in ferocity as he recedes from the usual hunting-fields.

At Oominak, one winter day, an Eskimo and his son were nearly killed by a bear that had housed himself in an iceberg. They attacked him with the lance, but he boldly turned on them, and handled them severely before they could make their escape.

The continued hostility of man, however, has had, in Dr. Kane's opinion, a modifying influence upon the ursine character in South Greenland; at all events, the bears of that region never attack, and even in self-defence seldom inflict injury upon, the hunters. Many instances have occurred where they have defended themselves, and even charged after having been wounded, but in none of them was life lost.

A stout Eskimo, an assistant to a Danish cooper of Upernavik, fired at a she-bear, and the animal closed at the instant of receiving the ball. The man had the presence of mind to fling himself prone on the ground, extending his arm to protect his head, and afterwards lying perfectly motionless. The beast was deceived. She gave the arm a bite or two, but finding her enemy did not stir, she retired a few paces, and sat upon her haunches to watch. But her watch was not as wary as it should have been, for the hunter dexterously reloaded his rifle, and slew her with the second shot.

It has been pointed out that in approaching the bear the hunters should take advantage of the cover afforded by the inequalities of the frozen surface, such as its ridges and hillocks. These vary in height, from ten feet to a hundred, and frequently are packed so closely together as to leave scarcely a yard of level surface. It is in such a region that the Polar bear exhibits his utmost speed, and in such a region his pursuit is attended with no slight difficulty.

And after the day's labour comes the night's rest; but what a night! We know what night is in these temperate climes, or in the genial southern lands; a night of stars, with a deep blue sky overspreading the happy earth like a dome of sapphire: a night of brightness and serene glory, when the moon is high in the heaven, and its soft radiance seems to touch tree and stream, hill and vale, with a tint of silver; a night of storm, when the clouds hang low and heavily, and the rain descends, and a wailing rushing wind loses itself in the recesses of the shuddering woods; we know what night is, in these temperate regions, under all its various aspects,—now mild and beautiful, now gloomy and sad, now grand and tempestuous; the long dark night of winter with its frosty airs, and its drooping shadows thrown back by the dead surface of the snow; the brief bright night of summer, which forms so short a pause between the evening of one day and the morning of another, that it seems intended only to afford the busy earth a breathing-time;—but we can form no idea of what an *Arctic Night* is, in all its mystery, magnificence, and wonder. Strange stars light up the heavens; the forms of earth are strange; all is unfamiliar, and almost unintelligible.

STALKING A BEAR.

It is not that the Arctic night makes a heavy demand on our physical faculties. Against its rigour man is able to defend himself; but it is less easy to provide against its strain on the moral and intellectual faculties. The darkness which clothes Nature for so long a period reveals to the senses of the European explorer what is virtually a new world, and the senses do not well adapt themselves to that world. The cheering influences of the rising sun, which invite to labour; the soothing influences of the evening twilight, which beguile to rest; that quick change from day to night, and night to day, which so lightens the burden of existence in our temperate clime to mind and soul and body, kindling the hope and renewing the courage,—all these are wanting in the Polar world, and man suffers and languishes accordingly. The grandeur of Nature, says Dr. Hayes, ceases to give delight to the dulled sympathies, and the heart longs continually for new associations, new hopes, new objects, new sources of interest and pleasure. The solitude is so dark and drear as to oppress the understanding; the imagination is haunted by the desolation which everywhere prevails; and the silence is so absolute as to become a terror.

The lover of Nature will, of course, find much that is attractive in the Arctic night; in the mysterious coruscations of the aurora, in the flow of the moonlight over the hills and icebergs, in the keen clearness of the starlight, in the sublimity of the mountains and the glaciers, in the awful wildness of the storms; but it must be owned that they speak a language which is rough, rugged, and severe.

All things seem built up on a colossal scale in the Arctic world. Colossal are those dark and tempest-beaten cliffs which oppose their grim rampart to the ceaseless roll and rush of the ice-clad waters. Colossal are those mountain-peaks which raise their crests, white with unnumbered winters, into the very heavens. Colossal are those huge ice-rivers, those glaciers, which, born long ago in the depths of the far-off valleys, have gradually moved their ponderous masses down to the ocean's brink. Colossal are those floating islands of ice, which, outrivalling the puny architecture of man, his temples, palaces, and pyramids, drift away into the wide waste of waters, as if abandoned by the Hand that called them into existence. Colossal is that vast sheet of frozen, frosty snow, shimmering with a crystalline lustre, which covers the icy plains for countless leagues, and stretches away, perhaps, to the very border of the sea that is supposed to encircle the unattained Pole.

In Dr. Hayes' account of his voyage of discovery towards the North Pole occurs a fine passage descriptive of the various phases of the Arctic night. "I have gone out often," he says, "into its darkness, and viewed Nature under different aspects. I have rejoiced with her in her strength, and communed with her in her repose. I have seen the wild burst of her anger, have watched her sportive play, and have beheld her robed in silence. I have walked abroad in the darkness when the winds were roaring through the hills and crashing over the plain. I have strolled along the beach when the only sound that broke the stillness was the dull creaking of the ice-floes, as they rose and fell lazily with the tide. I have wandered far out upon the frozen sea, and listened to the voice of the icebergs bewailing their imprisonment; along the glacier, where forms and falls the avalanche; upon the hill-top, where the drifting snow, coursing over the rocks, sung its plaintive song; and again, I have wandered away to some distant valley where all these sounds were hushed, and the air was still and solemn as the tomb."

Whoever has been overtaken by a winter night, when crossing some snowy plain, or making his way over the hills and through the valleys, in the deep drifts, and with the icicles pendent from the leafless boughs, and the white mantle overspreading every object dimly discernible in the darkness, will have felt the awe and mystery of the *silence* that then and there prevails. Both the sky above and the earth beneath reveal only an endless and unfathomable quiet. This, too, is the peculiar characteristic of the Arctic night. Evidence there is none of life or motion. No footfall of living thing breaks on the longing ear. No cry of bird enlivens the scene; there is no tree, among the branches of which the wind may sigh and moan. And hence it is that one who had travelled much, and seen many dangers, and witnessed Nature in many phases, was led to say that he had seen no expression on the face of Nature so filled with terror as the silence of the Arctic night.

But by degrees the darkness grows less intense, and the coming of the day is announced by the prevalence of a kind of twilight, which increases more and more rapidly as winter passes into spring. There are signs that Nature is awakening once more to life and motion. The foxes come out upon the hill-side, both blue and white, and gallop hither and thither in search of food, —following in the track of the bear, to feed on the refuse which the "tiger of the ice" throws aside. The walrus and the seal come more frequently to land; and the latter begins to assemble on the ice-floes, and select its breeding-places. At length, early in February, broad daylight comes at noon, and then the weary explorer rejoices to know that the end is near. Flocks of speckled birds arrive, and shelter themselves under the lee of the shore; chiefly *dove-kies*, as they are called in Southern Greenland—the *Uria grylle* of the naturalist. At last, on the 18th or 19th of February, the sun once more makes its appearance above the southern horizon, and is welcomed as one welcomes a friend who has been long lost, and is found again. Upon the crests of the hills light clouds are floating lazily, and through these the glorious orb is pouring a stream of golden fire, and all the southern sky quivers, as it were, with the shooting, shifting splendours of the coming day. Presently a soft bright ray breaks through the vaporous haze, kindling it into a purple sea, and touches the silvery summits of the lofty icebergs until they seem like domes and pinnacles of flame. Nearer and nearer comes that auspicious ray, and widens as it comes; and that purple sea enlarges in every direction; and those domes and pinnacles of flame multiply in quick succession as they feel the passage of the quickening light; and the dark red cliffs are warmed with an indescribable glow; and a mysterious change passes over the face of the ocean; and all Nature acknowledges the presence of the sun!

"The parent of light and life everywhere," says Dr. Hayes, "he is the same within these solitudes. The germ awaits him here as in the Orient; but there it rests only through the short hours of a summer night, while here it reposes for months under a sheet of snows. But after a while the bright sun will tear this sheet asunder, and will tumble it in gushing fountains to the sea, and will kiss the cold earth, and give it warmth and life; and the flowers will bud and bloom, and will turn their tiny faces smilingly and gratefully up to him, as he wanders over these ancient hills in the long summer. The very glaciers will weep tears of joy at his coming. The ice will loose its iron grip upon the waters, and will let the wild waves play in freedom. The reindeer will skip gleefully over the mountains to welcome his return, and will look longingly to him for the green pastures. The sea-fowls, knowing that he will give them a resting-place for their feet on the rocky islands, will come to seek the moss-beds which he spreads for their nests; and the sparrows will come on his life-giving rays, and will sing their love-songs through the endless day."

With the sun return the Arctic birds, and before we quit the realm of waters we propose to glance at a few of those which frequent the cliffs and shores during the brief Polar summer.

Among the first-comers is the dove-kie or black guillemot (*Uria grylle*), which migrates to the temperate climates on the approach of winter, visiting Labrador, Norway, Scotland, and even descending as far south as Yorkshire. In fact, we know of no better place where to observe its habits than along the immense range of perpendicular cliffs stretching from Flamborough Head to Filey Bay. Here, on the bare ledges of this colossal ocean-wall, the guillemot lays its eggs, but without the protection of a nest; some of them parallel with the edge of the shelf, others nearly so, and others with their blunt and sharp ends indiscriminately pointing to the sea. They are not affixed to the rock by any glutinous matter, or any foreign substance whatever. You may see as many as nine or ten, or sometimes twelve, old guillemots in a line, so near to each other that their wings almost touch. The eggs vary greatly in size and shape and colour. Some are large, others small; some exceedingly sharp at one end, others rotund and globular. It is said that, if undisturbed, the guillemot never lays more than one egg; but if that be taken away, she will lay another, and so on. But Audubon asserts that he has seen these birds sitting on as many as three eggs at a time.

SEA-BIRDS IN THE POLAR REGIONS.

The black guillemot differs from the foolish guillemot (*Uria troile*) only in the colour of its plumage, which, with the exception of a large white patch on the coverts of each wing, is black, silky, and glossy; the feathers appearing to be all unwebbed, like silky filaments or fine hair. The bill, in all the species, is slender, strong, and pointed; the upper mandible bending slightly near the end, and the base covered with soft short feathers. The food of the guillemot consists of fish and other marine products.

The *Alcidæ*, or auks, are also included amongst the Arctic birds. The little auk (*Arctica alca*) frequents the countries stretching northwards from our latitudes to the regions of perpetual ice, and is found in the Polar Regions both of the Old World and the New. Here, indeed, they congregate in almost innumerable flocks. At early morn they sally forth to get their breakfast, which consists of different varieties of marine invertebrates, chiefly crustaceans, with which the Arctic waters teem. Then they return to the shore in immense swarms. It would be impossible, says an Arctic voyager, to convey an adequate idea of the numbers of these birds which swarmed around him. The slope on both sides of the valley in which he had pitched his camp rose at an angle of about forty-five degrees to a distance of from 300 to 500 feet, where it met the cliffs, which stood about 700 feet higher. These hill-sides are composed of the loose rocks detached from the cliffs by the action of the frost. The birds crawl among these rocks, winding far in through narrow places, and there deposit their eggs and hatch their young, secure from their great enemy, the Arctic fox.

THE AUK.

On one occasion, they were congregated along a slope, fully a mile in length, and over this slope rushed a constant stream of birds, only a few feet above the stones; and, after making in their rapid flight the whole length of the hill, they returned higher in the air, performing over and over again the complete circuit. Occasionally a few hundreds or thousands of them would drop down, as if following some leader; and in an instant the rocks, for a space of several rods, would swarm all over with them, their black backs and pure white breasts speckling the hill very prettily.

Though quantities are destroyed by the crews of vessels as well as by the Eskimos, their numbers never seem to decrease. Their flesh is both wholesome and delicate, and affords a welcome change of diet to the mariner weary of salt meat and pemmican. They are very tame, and easily captured,—in some places being actually caught in hand-nets, like moths or butterflies; and they pass a great portion of their time on the ocean, where they disport themselves with equal grace and self-possession.

The starakis (*Phaleridinæ*) inhabit the archipelagoes which lie between China and North America. They assemble in small flocks, and swim about in quest of the crustaceans, molluscs, and other marine animals on which they feed. At nightfall they return to land, where they find shelter under the ledges of the rocks, or in burrows dug with their bill and feet. The female lays a solitary egg.

The auks abound in the high northern latitudes. They are all ocean-birds, and are never found, like the divers, in fresh-water streams and lakes. Those species which possess the power of flight nestle on the rocky cliffs and icebergs, where they lay a single egg, of conical form; a shape which prevents it from rolling away, or moving, except within a very narrow circle, on the bare rocky ledge where it is deposited.

The puffins (*Fratercola*), which in winter abound on our own shores, live chiefly on the water. They dive and swim with dexterity, but, owing to the shortness of their wings, are capable only of limited flight. Their plumage is thick, smooth, and dense, and so completely throws off the water that it is quite impervious to wet; while their deep, compressed, and pointed beak, resembling exactly a double keel, is admirably adapted as an instrument for cutting the waves when the bird wishes to dive.

The puffins live principally upon sprats and other small fishes; and the food intended for their young they retain until partially digested, when they disgorge it into their mouths. Like all the auks, the mother-bird lays but one egg.

The appearance of an island or iceberg frequented by these birds is very vividly sketched by Audubon, than whom no naturalist has ever more completely attained a thorough acquaintance with the Bird-World.

He tells us that on every crag or stone stood a puffin, at the entrance of every hole another, and yet the sea was covered and the air filled with them. The burrows were all inhabited by young birds, of different ages and sizes; and clouds of puffins flew over us, each individual hold-

PUFFINS.

ing a small fish by the head. The burrows all communicated with each other in various ways, so that the whole island seemed to be perforated by a multitude of subterranean labyrinths, over which it was impossible to run without the risk of falling at almost every step. The voices of the young sounded beneath the traveller's foot like voices from the grave, and the stench was exceedingly disagreeable.

Something must next be said of the mergansers (*Merginæ*), a sub-family of the palmipeds, which also belong to the Polar world. Their principal characters may thus be stated: a straight bill, much compressed on the sides, and convex towards the tip, which is furnished with a broad and much-hooked nail; the wings are moderate, and pointed; the tail is short and rounded; the tarsi are short, and the toes moderate, the outer being as long as the middle, the three anterior ones united by a full web, while the hind toe is moderate, elevated, and provided with a broad web on its margin.

From these characters it is easy to infer that the bird is aquatic in its habits; that it can swim and dive well; that it is also capable of strong, swift flight; and that its food will consist chiefly of fishes.

The dun diver or goosander (*Mergus merganser*) is widely distributed throughout the Polar Regions both of the eastern and western continents. During its southern migration, it visits the United States, as well as France, Holland, and Germany; but on the approach of summer it retires to Siberia and Kamtschatka, Iceland, Greenland, and the Arctic shores of North America.

In these localities it constructs its nest—always near the edge of the water; building it up of grass, roots, and similar materials, with little regard to symmetry, and lining it with down. It is placed sometimes among the mossy, weedy stones; and sometimes it is concealed in the long grass, or under the cover of bushes, or in the stumps or hollows of decayed trees. The female lays from twelve to fourteen eggs, of a cream-yellow colour; their form is a long oval, both ends being equally obtuse. The goosander may be said to spend its time in the air and on the water; and, in truth, on the land it moves but laboriously and awkwardly, owing to the backward position of its legs. It rises with difficulty from the ground; but when once on the wing, its course is swift, strong, and steady. As it lives mainly upon fish, its flesh is oily and ill-flavoured; a circumstance which goes far to compensate the sportsman for the frequent failure of his attempts to capture it. It is a wild and wary bird, and as it swims with rapidity and dives with ease, it generally effects its escape from all but the most experienced hunters.

THE GOSSANDER.

Another species which abounds in northern latitudes is the smew (*Mergus albellus*), also known as the white nun or white merganser. This palmiped is about the size of a widgeon; is of elegant form; and its plumage beautifully coloured with black and white. Its bill is of a dusky blue, nearly two inches long, thickest at the base, and tapering into a slenderer and more narrow shape towards the point. An oval black patch, glossed with green, marks each side of the head; the under part of the crest is black; but all the rest of the head and neck, as well as the graceful breast and the belly, are white as snow, with the exception of a curved black line on each side of the upper part of the breast, and similar marks on the lower part; the back, the coverts on the ridge of the wings, and the primary quills are black; the secondaries and greater coverts are white-tipped; while the sides of the body, under the wings to the tail, exhibit a curious variegation of dark wavy lines. The legs and feet are of a leaden blue.

The range of the smew is very extensive, for it migrates as far southward as the Mediterranean, while it is found everywhere in the Arctic Regions.

On the shores of Novaia Zemlaia, as on those of Spitzbergen, the sea-birds arrive in countless hosts as soon as the summer sun has removed the long and dreary spell under which Nature labours through the winter months. The narrow rock-ledges on which they congregate, and where auks and guillemots assemble in thousands, the Russians call "a bazaar." The large gray

sea-mew (*Larus glaucus*), the "burgomaster" of the Dutch whalers, prefers the lonely summits of isolated cliffs, where it can reign the monarch of all it surveys. The ivory gull (*Larus eburneus*) is seldom found in high northern latitudes; but the common gull (*Larus canus*) and the black-backed gull (*Larus marinus*) are almost as abundant as guillemots.

THE BLACK-BACKED GULL.

In Iceland one of the most useful, and certainly not the least beautiful, of the birds is the eider-duck (*Somateria mollissima*), which also frequents the shores of Baffin and Hudson Bays, Lapland, Greenland, and Spitzbergen. It loves to breed on the small flat islands which lie off the coast, such as Akeney, Flutry, and Videy, where it is secure from the attacks of the Arctic fox. Its breeding-places in Iceland are private property, and some of them have been for centuries in the possession of the same families, which owe to the birds all their wealth and prosperity. Hence they are very vigilantly guarded. Whoever kills one is fined thirty dollars; and to secrete an egg, or pocket a few downs, is an offence punishable by law. The chief occupation of some of the proprietors is to examine through their telescopes all the boats that approach, so as to be sure that there are no guns on board.

THE EIDER-DUCK.

As the birds on these islands are quite tame, the eider-down is easily collected. The female having laid five or six pale greenish-olive eggs, in a nest fashioned with marine plants, and thickly lined with down of the most exquisite delicacy, the collectors carefully remove her, rob the nest of its precious lining, and then replace the bird. Immediately she begins to lay afresh, and again has recourse to the down on her body to protect her eggs; and should her own stock be exhausted, as is not unfrequently the case, she is furnished with an auxiliary supply by the male. Even this second lining is often taken away, and the poor bird a third time repeats the process, both as regards the eggs and the down; but if the plunderers do not spare her now, she afterwards abandons the nest, and seeks a home in some more sequestered nook.

As it comes to the European markets, this down, which is highly valued on account of its lightness, elasticity, and warmth, occurs in balls about the size of a man's fist, and weighing from three to four pounds. Such is its fineness and elastic quality, that when a ball is opened, and cautiously laid near the fire to expand, it will completely fill a quilt five feet square. It should be noted, however, that the down from dead birds is of comparatively little value, having lost its elasticity.

An interesting account of a visit to Vigr in the Isafjardardjufs, a favourite resort of the eider-duck in the north of Iceland, is furnished by Mr. Shepherd :—

As he approached the island, he says, he could see flocks upon flocks of the sacred birds, and could hear their cooings at a great distance. Landing on a rocky wave-worn shore, against which the waters scarcely rippled, he set off to survey the island. The shore he describes as "the most wonderful ornithological sight imaginable." The ducks and their nests were everywhere. Great brown ducks started up under his feet at every step ; and it was with difficulty that he avoided treading on some of the nests. As the island is but three-quarters of a mile across, the opposite shore is soon reached. On the coast was a wall built up of large stones, just above the high water-mark, about three feet high, and of considerable thickness. At the bottom, on both sides of it, alternate stones had been left out, forming a series of square compartments in which the ducks might make their nests. Almost every compartment was occupied ; and as the human intruder walked along the shore, a long line of startled ducks flew out one after the other. The surface of the water also was white with ducks, who welcomed their "brown wives" with loud and clamorous cooing.

Mr. Shepherd, on arriving at the farmhouse, was received in the most hospitable manner, hospitality being one of the special virtues of the Icelander. He was much impressed by the appearance of the house, which seemed to be converted into one large *duckery*. The earthen wall surrounding it, and the window-embrasures, were filled with ducks; on the ground, encircling the house, was a ring of ducks ; on the sloping roof were seated ducks ; and a duck was perched on the door-scraper !

A grassy bank close by had been cut into square patches like a chess-board (a square of turf of about eighteen inches being removed, and a hollow excavated), and all these squares were occupied by ducks. A windmill was infested with them, and so were all the out-houses, mounds, rocks, and crevices. In fact, the ducks were everywhere. Many of them were so tame as to allow the stranger to stroke them on their nests ; and their mistress said there was scarcely a duck on the island which would not allow her to take its eggs without flight or fear. When she first became possessor of the island, the produce of down from the ducks did not exceed fifteen pounds weight in the year, but under her careful nurture it had risen, in twenty years, to nearly one hundred pounds annually. About a pound and a half are required to make a coverlet for a single bed ; and the down is worth from twelve to fifteen shillings per pound. Most of the eggs are taken and pickled for winter consumption, one or two only being left to hatch.

Eider-ducks congregate in numerous flocks, generally in deep water; they dive with wonderful force, and thus are enabled to capture the shell-fish which form their principal food. If a storm threatens, they retire to the rocky shores where they love to breed and rest. The Greenlanders kill them with darts, pursuing them in their little boats, watching their course by the air-bubbles that come floating upward when they dive, and dexterously aiming at them as soon as they rise to the surface wearied. Their flesh is eaten by the Greenlanders, but it is not well-flavoured ; their eggs, however, are held in high esteem.

The king eider (*Somateria spectabilis*) belongs to the same genus as the former.

We suppose that every reader is acquainted with the beautiful lines in which Tennyson has embodied the fable of the dying swan singing its own dirge:—

"With an inner voice the river ran,
Adown it floated a dying swan,
And loudly did lament......
The wild swan's death-hymn took the soul
Of that waste place with joy
Hidden in sorrow: at first, to the ear
The warble was low, and full, and clear;......
But anon her awful jubilant voice,
With a music strange and manifold,
Flowed forth on a carol free and bold......
And the creeping mosses and clambering weeds,
And the willow-branches hoar and dank,
And the wavy swell of the soughing reeds,
And the wave-worn horns of the echoing bank,
And the silvery marish-flowers that throng
The desolate creeks and pools among,
Were flooded over with eddying song."

But the wild swan's voice, even in its death-hour, has no such musical sweetness as the poet here sets forth. It is always harsh and dissonant, and when it breaks on the silence of the Arctic skies carries with it an almost painful impression.

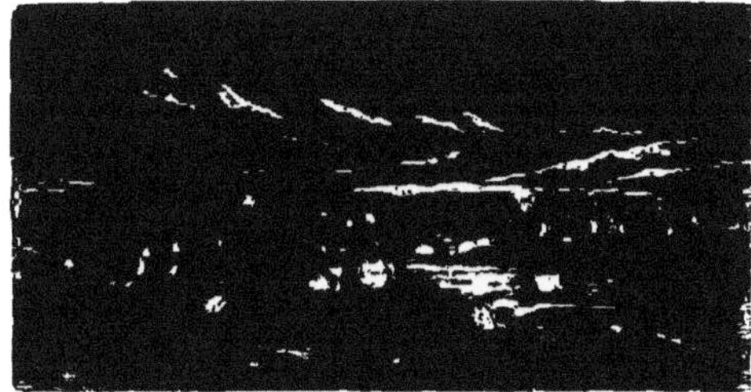

THE HAUNT OF THE WILD SWAN.

The lakes of Iceland, and its streams, abound with these beautiful birds. They are very numerous on the Myvatn, or Great Lake, as well as the wild duck, the scoter, the common goosander, the red-breasted merganser, the scaup duck, and other anserines. The wild swan is shot or caught for its feathers, which are highly prized for ornamental purposes. It is sometimes found in large flocks, sometimes in single pairs; and besides the lakes and streams, it frequents the salt and brackish waters along the coast. It is chiefly at the pairing season, or at the approach of winter, that it assembles in multitudes; and as the winter advances it mounts high in air, and shapes its course in search of milder climates.

The female builds her nest of the withered leaves and stalks of reeds and rushes, in lonely and sequestered places. She usually lays six or seven thick-shelled eggs, which are hatched in about six weeks, when both parents assiduously guard and feed the cygnets. When full-grown, this fine bird measures nearly five feet in length, and above seven in breadth across its extended wings; it weighs about fifteen pounds. The entire plumage is of a pure white, and next to the skin lies a coat of thick fine down.

The wealth of the Arctic and sub-Arctic seas is apparently inexhaustible. In many parts cod are plentiful, and supply the Greenlanders with a valuable article of food. The capelin (*Mullotus vitlosus*), which in May and June frequents the Greenland waters, is eaten both fresh and dried; in the latter case forming a useful winter provision. The halibut is found of a large size; and ocean also contributes the Norway haddock, the salmon-trout, the lump-fish, and the bull-head. Nor are the crustacea unrepresented: long-tailed crabs being abundant, while the common mussel may be gathered almost everywhere at ebb-tide. The seas, however, grow poorer as we advance towards the Pole, and many important species of fish do not penetrate further north than the Arctic Circle.

Yet even where these are wanting, the ocean-waters teem with life; and a recent writer is fully justified in remarking that the vast multitudes of animated beings which people them form a remarkable contrast to the nakedness of their bleak and desolate shores. The colder surface-waters are, as he says, almost perpetually exposed to a cold atmosphere, and being frequently covered, even in summer, with floating ice, they are not favourable to the development of organic life; but this adverse influence is modified by the higher temperature which constantly prevails at a greater depth. Contrary to the rule in the Equatorial seas, we find in the Polar ocean an increase of temperature from the surface downwards, in consequence of the warmer under-currents, flowing from the south northwards, and passing beneath the cold waters of the superficial Arctic current.

Hence the awful rigour of the Arctic winter, which strikes the earth with a death-blight, is not perceptible in the ocean-depths, where myriads of organisms find a secure retreat from the frost, and whence they emerge during the long summer's day, either to haunt the shores or ascend the broad rivers of the Polar world. Between the parallels of 74° and 80°, Dr. Scoresby observed that the colour of the Greenland sea varies from the purest ultramarine to olive-green, and from crystalline transparency to striking opacity; and these appearances are not transitory, but permanent.* The aspect of this green semi-opaque water, which varies in its locality with the currents,—often forming isolated stripes, and sometimes spreading over two or three degrees of latitude,—is mainly due to small medusæ and nudibranchiate molluscs. Many thousands of square miles must literally run riot with life, since the coloured waters we speak of are calculated to form one-fourth of the sea between the 74th and 80th parallels.

On the Greenland coast, where the transparency of the waters is so great that the bottom and every object upon it are clearly discernible, even at a depth of eighty fathoms, the ocean-bed is covered with gigantic tangles, so as to remind the spectator of the ocean-gardens of the Tropical Zone. Alcyonians, sertularians, ascidians, nullipores, mussels, and a variety of other sessile animals incrust every stone, or congregate in every fissure and hollow of the rocky ground. A dead seal or fish flung into the sea is soon converted into a skeleton, it is said, by the myriads

* Scoresby calculated that it would require 80,000 persons, labouring continuously from the creation of man to the present day, to count the number of organisms contained in two miles of the green water.

of small crustaceans which infest these northern waters, and, like the ants in the equatorial forests, perform the part of scavengers of the deep.

It is evident, from the observations of Professor Forbes, that *depth* has a very considerable influence in the distribution of marine life. From the surface to the depth of 1380 feet eight distinct zones or regions have been mapped out in the sea, each of which has its own vegetation and inhabitants; and the number of these regions must now be increased, after the astonishing results of the deep-sea soundings of Dr. Carpenter and Professor Wyville Thomson. The changes in the different zones are not abrupt: some of the creatures of an under region always appear before those of the region above it vanish; and though there are a few species the same in some of the eight zones, only two are common to all. It is to be observed that those near the surface have forms and colours analogous to the inhabitants of southern latitudes, while those at a greater depth are analogous to the animals of northern waters. Hence, in the sea, *depth* corresponds with latitude, as *height* does on land. Mrs. Somerville adds, in language of much terseness, that the extent of the geographical distribution of any species is proportioned to the depth at which it lives. Consequently, those which live near the surface are less widely dispersed than those inhabiting deep water.

The larger and more active inhabitants of the seas obey the same laws with the rest of creation, though their provinces, or regions, are in some instances very extensive. Above the 44th parallel the Atlantic species frequently correspond with those of the Pacific. The salmon of America is identical with that of the British Isles, and the coasts of Sweden and Norway; the same is true of the *Gadidæ*, or cod. The *Cottus*, or bull-head tribe, are also the same on both sides of the Atlantic; increasing in numbers and specific differences on approaching the Arctic seas. The same law holds good in the North Pacific, but the *generic* forms differ from those in the Atlantic. From the propinquity of the coasts of America and Asia at Behring Strait, the fish on both sides are nearly alike, down to Admiralty Inlet on the one side, and the Sea of Okhotsk on the other.

CHAPTER IV.

THE GLACIERS.

S introductory to a description of the Arctic Glaciers, a few words on the formation of snow seem necessary. Briefly, it may be said that snow is the result of the crystallization of water.

The molecules and atoms of all substances, when not constrained by some external power, build themselves up into crystals. This is true of the metals and minerals, if, after having been melted, they are allowed to cool gradually. Bismuth develops the process in a very impressive manner, and when properly fused and solidified exhibits large-sized crystals of singular beauty.

In like manner, sugar dissolved in water produces, after evaporation has taken place, crystals of sugar-candy. The ready crystallization of alum is known to every school-boy who has dabbled in "chemical experiments." Chalk dissolved and crystallized becomes Iceland spar, and assumes a variety of fanciful and graceful shapes. The diamond is crystallized carbon; and the crystallizing power is inherent in all our precious stones,—sapphire, topaz, emerald, beryl, amethyst, ruby.

In the process of crystallization, it is found that the minutest particle of matter is possessed of an attractive and a repellent pole, and that by their natural action the form and structure of the crystal are determined.

The attracting poles, in the solid condition of any given substance, are firmly interlocked; but dissolve the cohesion by the application of sufficient heat, and the poles will recede so far as to be practically beyond each other's range. And thus the natural tendency of the molecules to build themselves together is neutralized.

Water, for example, as a liquid is, to all appearance, without form; but when sufficiently cooled, its molecules are brought under the influence of the crystallizing force, and then arrange themselves in the most varied and beautiful shapes. When snow falls in calm air, the icy particles present themselves in the form of six-rayed stars. From this type there is no departure, though the appearance of the snow-stars in other respects is infinitely varied.

It is worth pausing, as Professor Tyndall remarks, to think what wonderful work is going on in the atmosphere during the formation and descent of every snow-shower: what "building power" is brought into play! and how imperfect seem the productions of human minds and hands when compared with those produced by the forces of Nature!

We have spoken of attracting and repelling poles; but a few words of explanation seem desirable. Every magnet possesses two such poles; and if iron filings be scattered over a magnet, each particle becomes also endowed with two poles. Now suppose that similar particles, devoid of weight, and floating in the atmosphere, come together, what will happen? Obviously, the repellent poles will retreat from each other, while the attractive will approach, and ultimately interlock. Further: if the particles, instead of a single pair, possess several pairs of poles arranged at definite points over their surfaces, you can then picture them, in obedience to their mutual attractions and repulsions, building themselves together in masses of definite shape and structure.

VARIOUS FORMS OF SNOW-CRYSTALS.

You have, then, only to imagine the aqueous particles in cold calm air to be gifted with poles of this description, compelling the said particles to assume certain definite aggregates, and you have before your mind's eye the invisible architecture which creates the visible and beautiful crystals of the snow.

The important part played by this crystallizing force in ice as well as snow, will be understood from the following remarks by Professor Tyndall, who may justly be described as the most eminent living authority on the subject:—

At any temperature below 32° F.,—that is, freezing-point,—the movement of heat is sufficient to loosen the molecules of water from their rigid bonds of cohesion. But at 32° the movement is so diminished that the atoms lock themselves together, and unite in a solid. This act of union, however, is controlled by well-known laws. To the unintelligent eye a block of ice seems neither more interesting nor more beautiful than a sheet of glass; but to the instructed mind the ice is to the glass what an oratorio of Handel is to the scream of a ballad-singer. Ice is music, glass is noise; ice represents order, glass confusion. In the latter, the molecular forces have brought about an inextricable intertangled network; in the former, they have woven a rich and regular embroidery, the designs of which are infinitely beautiful.

Let us suppose ourselves examining a block of ice. In what way shall we get at its structure? A sunbeam, or if that be wanting, a ray of electric light is the anatomist to which we must confide the work of dissection. We direct this ray straight from our lamp across the plate of transparent ice.

It shivers into pieces the icy edifice, exactly reversing the order of its architecture.

The crystallizing force, for example, had silently and systematically built up atom after atom; the electric ray dislocates them (so to speak) just as silently and systematically.

We elevate the ice-block in front of the lamp, so that the light may now pass through its substance. Compare the ray as it enters with the ray as it makes its exit; to the eye there is no perceptible difference, and its intensity seems scarcely diminished. But not so with its heat. As a thermic agent, the ray was more powerful before its entrance than it was after its emergence. A portion of its heat is arrested, is detained in the ice, and of this portion we now proceed to avail ourselves. What will it effect?

We place a lens in front of the ice upon the screen. Now, observe this image (see Illustration), the beauty of which is still very far from the real effect. Here is one star; yonder is another; and in proportion as the action continues, the ice appears to resolve itself more and more into stars, all of six rays, like snow-crystals, and resembling a beautiful flower. By moving the lens in and out, we bring new stars into sight; and while the action continues, the edge of the petals is covered with indentations like those of the leaf of a fern. Probably, few of our readers have any conception of the magical beauties concealed in a block of ice! Let them remember that prodigal Nature works in this way throughout the whole world. Every atom of the solid crust which covers the frozen waters of the North, has been wrought out in obedience to the law we have enunciated. Nature is always and everywhere harmonious; and it is the mission of Science to awaken us to an appreciation of its concords.

EXHIBITION OF ICE-FLOWERS BY PROJECTION.

ICE-FLOWERS.

There is another point of our experiment to which the reader's attention must be directed. He sees the flowers illuminated by the ray which traverses them. But if he examines them, while turning upon them a ray which they will reflect and send back to his own eye, he will see in the centre of each a spot with the brightness of burnished silver. He will be tempted to think that this spot is a bubble of air; but, by immersing the ice in hot water, you can melt the ice all around the spot,—and when it alone remains, you will see it diminish and disappear without any trace of air. The spot is a vacuum. Such is the faithfulness to herself with which Nature operates; thus, in all her operations, does she submit to her own laws. We know that ice, in melting, contracts; and here we arrest the contraction, as it were, in the very act. The water of the flowers cannot fill the space occupied by the ice which by its fusion has given birth to them; hence the production of a vacuum, the inseparable companion of each liquid flower.

The fragment of compact ice whose elements assume such beautiful crystalline forms is itself a crystal. This was shown by Sir David Brewster, who employed for the purpose of analysis that modified form of light which we call *polarised light*. It is singularly well adapted to bring out the peculiarities of the main structure of substances, owing to the coloured figures which it outlines on a screen after passing through them. All crystals with an axis—such, for instance, as Iceland spar—yield a series of brilliantly-tinted rings, traversed by a regularly-formed cross entirely black. As ice produces the same figures, we are justified in attributing to it the same kind of crystallization. We must note, however, that we are referring now to the thick ice formed on our canals and lakes. If we examined the first film formed on the surface of the water, we should discover in it a completely irregular crystallization, the ray of polarised light producing only a mosaic of varied tints, distributed without any order. But it is easy to explain the way in which this primary crust or film is produced. Those portions of the fluid mass in contact with the air are the first to freeze, but each molecule of ice abandons its heat to the contiguous water, which thereby is slightly raised in temperature, and the result is a partial congelation. The surface we are examining then presents a network of fine needles intercrossed in every direction, and forming a kind of delicate lace, the meshes or intervals of which are gradually filled up. When the network is transformed into a continuous sheet, the loss of heat is diminished more and more as this external crust grows thicker and thicker; but the development of the ice invariably takes place by means of long interlaced needles, as the reader may see for himself by breaking off a portion from the nearest pond (in winter), and examining the sectional surface.

Having said thus much in reference to the crystallization of ice and snow, we proceed to explain the *regelation* and *moulding of ice*. Some years ago, Faraday astonished the scientific world by a very curious experiment. Splitting into two parts a piece of ice, he brought together the parts at the moment that fusion took place on their surfaces, and they united immediately. How are we to account for this effect, which can be produced even in hot water?

When the temperature of water rises, the surface molecules first become liquid, then gaseous; being placed beyond the coercitive action of the surrounding particles, they are easily set free; transported, on the contrary, into the centre of the mass, they are brought absolutely under the influence of this action, which induces a new solidification,—or, to use the scientific term, a *regelation*. In this way it becomes easy to understand how very various forms can be communicated by simple pressure to a fragment of ice. If the observer successively places a straight bar in moulds of increasing curvature, he may easily compel it to assume the shape of a ring or even of a knot. In each mould, it is true, the ice breaks; but if the pressure is kept up, the surfaces of the fragments are brought into contact, and adhere so as to re-establish a condition of continuity. A snowball may thus be converted into a sphere of ice, and the sphere, by constant pressure, into a cup or a statue.

Professor Tyndall refers to a remarkable instance of regelation which he observed one day in early spring. A layer of snow, not quite two inches thick, had fallen on the glass roof of a small conservatory, and the internal air, warming the panes, had melted the snow so far as it was in immediate contact with them. The entire layer had slipped down the pane, and projected beyond the edge of the roof, without falling, and had bent and curved as required, just like a flexible body.

MOULDING ICE.

The snow-fields which overspread the upper part of every glacier, whether in the Arctic Regions or elsewhere, are composed of crystallized snow, whose fragile, delicate, and fairy-like architecture endures so long as it remains dry, but undergoes a great transformation when the sun, melting the upper stratum, allows the water to interpenetrate its substance. The fluid, congealing anew during the night, transforms the snow into the condition technically known as *névé;* a term given by the Swiss physicists to a granular mass composed of small rounded icicles, disaggregated, but more adhesive than snow-flakes, and of a density intermediate between that of snow and that of ice. Under the pressure of new layers, and as a result of infiltrations of water, the *névé* unites, and solders into ice of constantly increasing compactness.

But glacier-ice presents some other curious peculiarities. Every abundant snowfall on the summit of the mountains forms a layer easily distinguishable from preceding layers—which, in most cases, have already passed into the *névé* condition. This stratification becomes more apparent when the whiteness of the surface has been sullied by dirt or dust wafted on "the wings of the wind." It is perceptible also in ice; but here we must not confound it with another phenomenon of which the cause is different, the *veined structure.*

In places where glaciers have been accidentally cut down in an almost vertical direction, the section is found to exhibit a series of parallel veins, formed by a beautiful and very transparent azure ice in the midst of the general mass, which is of a whitish colour, and slightly opaque.

In different glaciers, and in different parts of the same glacier, these blue veins will vary in number and intensity of colouring. They are specially beautiful in crevasses of recent formation, and on the sides of channels excavated in the ice by tiny rills resulting from superficial fusion. Not a few glaciers exhibit this remarkable veined structure throughout their entire extent. When a vertical cutting exposes the delicate azure network to atmospheric influences, the softer ice melts prior to the fusion of the blue ice which then remains in their detached leaflets. On examining these attentively, we cannot fail to remark the absence, or, at all events, the extreme rarity, of air-bubbles, though they are so plentiful in the coarser ice.

Professor Tyndall's explanation of this phenomenon is as interesting as it is ingenious. While on a visit of inspection to the slate-quarries of Wales, he had occasion to study the *cleavage* of the rocks which compose them; in other words, their faculty of dividing naturally, a property inherent in all crystals. The schistous slate separates easily into sheets, and in traversing different quarries one sees that all the planes of cleavage are parallel in each. From this circumstance our men of science were at first induced to look upon slates as the products of the stratification of different deposits. Such an explanation, however, could not be accepted by Tyndall, when he observed that the minute fossils embedded in them were constantly misshapen and flattened in the direction of the plane of cleavage, because the great modification they had undergone could not have taken place in superimposed strata at the bottom of the primeval sea. He concluded that these schists, therefore, must have been subjected to a considerable pressure; and further, that this pressure must have been exercised at right angles with the plane of separation of the different layers.

A long series of experiments proved that many bodies, when forcibly compressed, exhibit in their structure a very distinctly marked lamination, and frequently veins of very great beauty.

He carefully examined iron which had passed under the steam-hammer, or through the rolling-mill; clay and wax were subjected to the hydraulic press. In all cases he detected signs of cleavage; and hence we are justified in the inference that the phenomenon is invariably produced by pressure in all bodies of irregular internal structure. Such is the result with glacier-ice, from whose mass the air-bubbles introduced by the snow are gradually expelled. At first of brilliant whiteness, it assumes, in the parallel layers corresponding to the planes of cleavage, those beautiful azure tints which characterize the veined structure. So little has it to do with stratification, that in places where this is apparent it has given rise to a series of horizontal lines, while the parallel veinings, in the same masses of ice, are all inclined at an angle of about 60°.

The tendency to cleavage in compact ice would seem to explain the regular form of those fragments or detached pieces with which some parts of the glaciers are covered. Usually they occur as cubes, or as rectangular parallelopipeds. The Alpine mountaineers name them *séracs*,—in allusion to their resemblance to certain cheeses which bear this name, and which are manufactured in rectangular boxes. They have been found in many parts of a really colossal size, measuring fifty feet in length, breadth, and depth, and as regular in shape as if they had been hewn with a chisel.

There are many interesting points connected with the formation and constitution of glaciers which we should gladly discuss, but we are confined by our limits to remarks of a general character, and we must now pass on to speak of the phenomena attendant upon their motion. No doubt, the traveller who for the first time comes in sight of one of these huge ice-rivers, and sees the mighty mass apparently rooted to its valley-bed, solid, unchangeable, adamantine, finds it hard to believe that it moves onward with a certain and an unresting, though a gradual progress. It looks like a noble river, suddenly petrified by some overwhelming force: congealed, as it flowed, in a moment, by some irresistible spell! Such, indeed, is the conception of the poet:—

"Ye ice-falls! ye that from the mountain's brow
Adown enormous ravines slope amain....
Torrents, methinks, that heard a mighty voice,
And stopped at once amid their maddest plunge!
Motionless torrents! silent cataracts!"

And this conception is justified by the *aspect* of the glacier. Thus, of the Glacier du Géant,

Professor Tyndall says:—"It stretches smoothly for a long distance, then becomes disturbed, and then changes to a great frozen cascade, down which the ice appears to tumble in wild confusion. Above the cascade you see an expanse of shining snow, occupying an area of some square miles." But we shall see that here, as in the world of man, appearances are deceitful, and that the glacier well deserves to be called an ice-river, in allusion to its regular and continuous motion.

Between the snow-fall in the higher regions of the globe, and the quantity of snow which every summer disappears through liquefaction, the difference is very considerable. The supply, so to speak, exceeds the demand, and a residuum is annually left. It is only below the perpetual snow-line that the snow created and accumulated in winter is wholly melted in the warm season. And, therefore, if for any considerable period the excess upon any particular mountain continued to accumulate, immense masses of ice would gradually rise to the extreme height in the atmosphere affected by aqueous phenomena.

Rendu, the Roman Catholic prelate, who first led the way to the discovery of the true nature of glaciers, says, very justly,—"The economy of the world would be soon destroyed, if at certain points accumulations of matter prevailed. The centre of gravity of the globe would be insensibly displaced, and the admirable regularity of its movements would be succeeded by disorder and perturbation. If the Poles did not send back to the Equatorial seas the waters which, reduced into vapour, issue daily from these burning regions, to be converted into ice in the Arctic and Antarctic Zones, ocean would be drained dry, and life would cease, as well as water, to circulate throughout our world. The Creator, however, in order to ensure the permanence of His almighty work, has called into existence the vast and powerful law of circulation, and this law the careful observer sees reproduced in all the economy of Nature. The water circulates from the ocean into the air, from the air it spreads over the earth, and from the earth it passes into the seas. The rivers return from whence they came, in order that they may issue forth anew; the air circulates around the globe, and, as it were, upon itself, passing and repassing successively at all the altitudes of the atmospheric column. The elements of every organic substance circulates in changing from the solid to the liquid or aëriform state, and in returning from the latter to the state of solidity or organization. It is not improbable that the universal agent which we designate under the name of fire, light, electricity, and magnetism, has probably also a circle of *circulation* as extensive as the universe. Should its movements ever be known to us more than they now are, it is probable that they would afford the solution of a host of problems which still defy the intellect of man. Circulation is the law of life, the method of action employed by Providence in the administration of the universe. In the insect, as in the plant, as in the human body, we find a circulation, or rather several circulations,—blood, humours, elements, fire, all which enter into the composition of the individual."

However fanciful may be some of the amiable prelate's speculations, it is certain that the glaciers obey this law of circulation. The snow-accumulations in the upper regions are to some extent reduced by the descent of the avalanches,—that is, of masses of snow and ice which detach themselves from the mountain-sides and dash headlong into the valleys below, where they are rapidly melted by the warmer atmosphere. But this would, in itself, be wholly insufficient. Another movement, at once more efficacious and more regular, is necessary; a movement which embraces the entire system of the ice-masses, and which carries the glaciers below the perpetual snow-line, so that every year they may give up a portion of their terminal extremities. The discovery of this general progression is one of the most fertile with which, of late years, the physics of the globe have been enriched.

Professor Tyndall rightly observes that there are numerous obvious indications of the existence of glacier-motion, though it is too slow to catch the eye at once. The crevasses change within certain limits from year to year, and sometimes from month to month; and this could not be if the ice did not move. Rocks and stones also are observed, which have been plainly torn from the mountain-sides. Blocks seen to fall from particular points are afterwards noticed lower down. On the moraines rocks are found of a totally different mineralogical character from those composing the mountains right and left; and in all such cases strata of the same character are found bordering the glacier higher up. Hence the conclusion that the foreign *boulders* have been *floated* down by the ice. Further, the ends or "snouts" of many glaciers act like ploughshares on the land in front of them, overturning with irresistible energy the huts and châlets that lie in their path. Facts like these have been long known to the inhabitants of the High Alps, who were thus made acquainted in a vague and general way with the motion of the glaciers. But Science cannot deal with generalities: it requires precise and accurate information; and this information, so far as the progression of the glaciers is concerned, has been obtained through the patient labours of Rendu, Charpentier, Agassiz, Desor, Vogt, Professor Forbes, Bravais, Charles Martins, Hopkins, Professor Tyndall, Colomb, John Ball, and Schlagintweit. Their experiments and observations have established the truth of certain immutable principles, and proved the existence of a general law of movement.

The accumulation of the débris hurled headlong by the mountains forms on the glacier-surface long lines of stone and earth, which are called *moraines;* these diverge in certain directions, according to the circumstances we now come to explain.

The landslips which occur on the banks or edges of the glacier give rise to the *lateral moraines,* which are enlarged and extended daily by the twofold effect of the fall of stones and débris, and the progressive movement which carries them along with the whole mass of ice. Towards the centre of the great glaciers, in almost every case, is found a *medial moraine;* the result of the encounter of the lateral moraines of two glaciers which have united into one. These superficial moraines participating in the movement of the glacier, each of their blocks eventually rolls to the foot of the terminal precipice, and thus a *frontal moraine* is formed on the very soil of the valley, like an embankment raised to prohibit the further advance of the ice. And, lastly, the bed of sand, gravel, pebbles, and detritus which is found beneath the glacier, and over which it glides, is called the *profound moraine.*

The furrows wrought by this last-named stratum on the bottom of the glacier-channels show the wonderful force of friction which the glacier exercises during its descent. The depths of these furrows depends entirely on the hardness of the débris carried down by the glacier, and the nature of the rocks submitted to the friction. The polish assumed by these rocks when they are sufficiently solid to resist the thunderous march of the glacier, indicates the enormous pressure which it exercises on the slopes of the valley through which it forces its way. This effort, bearing principally on the side of the rocks turned in the direction of their crests, impresses upon them a peculiar rounded form, so like the appearance of a flock of sheep (*moutons*) that De Saussure gave them the name of *roches moutonnées*.

Connected with the scientific evidence of the progressive movement of glaciers, a glacier in the Bernese Oberland will for ever be memorable. Two branch glaciers, the Lauteraar and the Finsteraar, unite at a promontory called the Abschwung to form the trunk-glacier of the Unteraar, which carries a great medial moraine along its colossal back.

Here in 1827, an "intrepid and enthusiastic" Swiss professor, Hugi, of Solothurn (or Soleure), erected a small cabin of stones for the purpose of observations upon the glacier. The hut moved, and he took steps to measure its motion. In three years, 1827 to 1830, it moved 330 feet downwards. In 1836 it had descended 2354 feet; and in 1841, it had accomplished a journey of 4712 feet. [This was at the rate of about 336 feet a year.]

In 1840, M. Agassiz, with some scientific friends, Messrs. Desor, Vogt, and Nicolieb, established themselves under a great overhanging slab of rock on the same moraine, and by means of side walls, and other appliances, constructed a rough abode which, because some of these men of science came from Neufchâtel, they named the "Hôtel des Neuchâtelois."

In two years after its erection, Agassiz discovered that it had moved downwards no less a distance than 486 feet.

These and some similar measurements brought to light a very important fact. The reader will observe that the *middle numbers*, corresponding to the central portion of the glacier, are the *largest:* hence it was obvious that the *centre of a glacier, like that of a river, moves more rapidly than the sides.*

Owing to the greater central motion of a glacier, its crevasses invariably assume a curved outline, of which the convexity advances towards the bottom of the valley.

It has also been ascertained that the superficial part of a glacier moves more rapidly than its base.

Again: Tyndall and Hirst, by employing instruments of great precision, have demonstrated that the maximum of motion is not to be found exactly in the centre, but that, according to the windings of the valley through which the glacier flows, it moves sometimes to the right of the centre, and sometimes to the left. Now, the progression of a river exhibits all the characters we have just enumerated, and the truth foreshadowed by Rendu has been confirmed in every detail. The glacier *is* a "river of ice."

The reader will naturally ask, How can a substance of such apparent rigidity as ice obey, as it *does* obey, the same laws which regulate the movement of fluids? I can understand, he may say, how water flows in such and such a manner: it is a liquid, and its molecules are deficient in the property of cohesion; but that so solid, and firm, and unimpressible a substance as ice should be capable of motion seems impossible. I can understand very easily that a mass of ice, when loosened or detached from its resting-place, will glide downwards until arrested by some adequate obstacle; but this is not the kind of motion you are describing. According to your explanations, every constituent portion of the glacier moves, and the central faster than the lateral, and the surface faster than the base.

These objections were advanced by men of science when the motion of glaciers was first put forward as a theory; and the answer given by Scheuchzer was, that a glacier might be compared, in the summer season, to a sponge saturated with water, which, when afterwards congealed by the cold temperature of autumn and winter, expanded, and produced a dilatation of the mass in every direction. Then, as it could not recede, as it could not *reascend* its valley-slope, the augmentation of size would necessarily take place in its lower portion.

It is unnecessary for us to explain why this answer was unsatisfactory. Subsequent observations, however, proved its impossibility, and Professor Forbes then put forward his ideas of the *viscous character* of ice. But these, too, did not meet the conditions of the phenomenon; and the view now adopted is that of Professor Tyndall, who has shown that it is the result of the regelation we have already described.

Professor Forbes enunciated his theory in words to the following effect: "A glacier is an imperfect fluid or viscous body, which is urged down slopes of certain inclination by the natural pressure of its parts." But we know the exceeding brittleness of ice, and how is viscosity compatible with brittleness? We know, too, that crevasses and fissures will suddenly form on a glacier, like the cracks on a pane of glass. But if ice were viscous, and could expand, dilate, or stretch as viscous substances do, these crevasses would be impossible. They would gradually close up, like an indent in a mass of jelly. And yet it cannot be denied that a glacier *does* move like a viscous body; the centre flowing past the sides, the top flowing over the bottom, while the motion through a curved valley corresponds to fluid motion. How are we to reconcile these apparently conflicting circumstances?

By Professor Tyndall's regelation theory, which is founded on a fact already mentioned; namely, that when two pieces of thawing ice are brought in contact, they freeze together.

This *fact*, and its application irrespective of the *cause* of regelation, may be thus illustrated: "Saw two slabs from a block of ice, and bring their flat surfaces into contact; they immediately freeze together. Two plates of ice, laid one upon the other, with flannel round them overnight, are sometimes so firmly frozen in the morning that they will rather break elsewhere than along their surface of junction. If you enter one of the dripping ice-caves of Switzerland, you have only to press for a moment a slab of ice against the roof of the cave to cause it to freeze there and stick to the roof.

"Place a number of fragments of ice in a basin of water, and cause them to touch each other; they freeze together where they touch. You can form a chain of such fragments; and then, by taking hold of one end of the chain, you can draw the whole series after it. Chains of icebergs are sometimes formed in this way in the Arctic seas."

From these observations we deduce the following result:—Snow consists of small particles of ice. Now, if by pressure we squeeze out the air entangled in thawing snow, and bring the little ice-granules into close contact, they may be expected, as they do, to freeze together; and should the expulsion of the air be complete, the squeezed snow will assume the appearance of compact ice.

It is in this way that the consolidation of the snows takes place in the Arctic as in the higher Alpine regions. The deeper layers of the *névé* are converted into more or less perfect ice by the pressure of the superjacent layers; and further, they are made to assume the shape of the valley which they fill, by the slow and continuous pressure of its sides.

In glaciers, as Professor Tyndall points out, we have ample illustrations of rude fracture and regelation; as, for example, in the opening and closing of crevasses. The glacier is broken on the cascades, and mended at their bases. When two branch glaciers lay their sides together, the regelation is so firm that they begin immediately to flow in the trunk glacier as in a single stream. The medial moraine gives no indication by its slowness of motion that it is derived from the sluggish ice of the sides of the branch glaciers.

We may sum up the regelation theory in few words. The ice of glaciers changes its form and retains its continuity under *pressure* which keeps its particles together. But when subjected to *tension*, sooner than stretch, it *breaks*, and behaves no longer as a viscous body.

These are Professor Tyndall's words, and the fact which they embody it would be difficult to set forth more clearly or more concisely.

A POLAR GLACIER.

Having said thus much of the structure, causes, characteristics, and movement of glaciers, we proceed to consider some of the more remarkable of those which are situated in the Arctic World.

The glaciers of the Polar Regions do not differ in structure or mode of formation from those of other countries. Yet they possess some peculiar features, and to a superficial observer might seem independent of the physical laws we have attempted to explain. That this is not the case has been shown by Charles Martins, who carefully studied the glaciers of Spitzbergen on the occasion of the exploring voyage of the *Recherche* to that island, and has demonstrated that their differences are but a particular case of the general phenomenon.

As special characters he points out, first, the rarity of needles and prisms of ice, which he attributes to the slight inclination and the uniformity of the slopes, as well as to the diminution of the solar heat, which, even in the long summer days, does not melt the surface. There are no rills or streams capable of hollowing out crevasses and moulding protuberances or projections. But transversal crevasses produced by the movement of the glaciers are numerous, and these are often very wide and very deep.

In the terminal escarpment, which melts in proportion as it plunges into the sea, immense caverns are sometimes seen; caverns so immense that the azure-gleaming grottoes of the Arveiron and Grindelwald, so much admired by European travellers, are but miniatures. "One day," says Charles Martins, "after having ascertained the temperature of the sea off the great glacier of Bell Sound, I proposed to the sailors who accompanied me to carry our boat into its cavern. I explained to them the risk we should incur, being unwilling to attempt anything without their consent. When our boat had crossed the threshold, we found ourselves in an immense Gothic

GLACIER, ENGLISH BAY, SPITZBERGEN.

cathedral; long conical-pointed cylinders of ice descended from the roof; the recesses seemed so many chapels opening out of the principal nave; broad fissures divided the walls, and the open intervals, like arches, sprang towards the summits; azure gleams played over the icy surface, and were reflected in the water. The sailors, like myself, were dumb with admiration. But a too prolonged contemplation would have been dangerous; we soon regained the narrow opening through which we had penetrated into this winter temple, and, returning on board our vessel, preserved a discreet silence respecting an escapade which might have been justly blamed. In the evening, we saw from the shore our cathedral of the morning slowly bend forwards, detach itself from the parent glacier, crash into the waves, and reappear in a thousand blocks and fragments of ice, which the retiring tide carried slowly out to sea."

The Spitzbergen glaciers do not exhibit those numerous moraines which are observed on the majority of those of Switzerland.

The mountains, not being very lofty, are buried, as it were, under their burden of glaciers, instead of preponderating over them, and seem with difficulty to lift their peaks out of the mass of ice and snow surrounding them. Consequently, there are no considerable landslips or falls of earth and stone, which, accumulating along the borders of the glaciers, might form moraines. Martins is of opinion that the Spitzbergen glaciers correspond to the upper part of the glaciers of Switzerland; to so much, that is to say, as lies above the perpetual snow-line.

GLACIER, BELL SOUND, SPITZBERGEN.

Now, he says, the higher we ascend on an Alpine glacier, the more do the lateral and medial moraines diminish in width and form, until they taper away and finally disappear under the high *névés* of the amphitheatres from which the glacier issues, just as the mountain torrents often take their rise in one or in several lakes terraced one above the other.

For all these reasons, he adds, the medial and lateral moraines are scarcely conspicuous on the glaciers of Spitzbergen; a number of stones and boulders may be seen along their sides, and sometimes in their centre, but the ice is never hidden, as in the Alps, under the mass of débris accumulated upon it. As for the terminal moraines, they must be sought at the bottom of the sea, since the terminal escarpment nearly always overhangs it. Hence, the blocks of stone fall simultaneously with the blocks of ice, and form a submarine frontal moraine, of which the two extremities are occasionally visible upon the shore.

In a previous chapter we have alluded to the manner in which icebergs are formed by the detachment from the seaward extremity of the glacier of huge masses of ice, which the current carries out into the open sea. To the description already given, we may here add that which Charles Martins furnishes in his valuable and interesting record of persevering scientific enterprise, "Du Spitzberg au Sahara":—In Spitzbergen, he says, the glacier, after a traject of more or less considerable duration, reaches the sea. If the shore be rectilineal, it advances no further; but, in the recess of a bay, where the shore is curved, it continues its progression, supporting its bulk on the sides of the bay, and advancing above the water, which it overhangs. This is easily understood. In summer the sea-water at the bottom of the bays is always at a temperature a little above 32°; on coming in contact with this comparatively warm water the glacier melts, and, at low tide, an interval is perceptible between the ice and the surface of the water. The glacier being no longer supported, partially crumbles and gives way; immense blocks detach themselves, fall into the sea, disappear beneath the water, reappear revolving on their own axes, and oscillate for a few moments until they have taken up their position of equilibrium. The blocks thus detached from the floating masses, of all sizes and shapes, are called icebergs.

Our traveller records that twice a day, in Magdalena Bay and Bell Sound, he was an eye-witness of this partial ruin of the extremity of the glaciers. Their fall was accompanied by a

STEAMER "CHARGING" AN ICEBERG, UPERNAVIK, GREENLAND

noise like that of thunder; the swollen sea rushed upon the shore in a succession of gigantic waves; the gulf was covered with icebergs, which, caught in the swirl and eddy, issued out of the bay, like immense fleets, to gain the sea beyond, or were stranded here and there at points where the water was shallow. The icebergs seen by M. Martins were not, however, of any surprising magnitude; he estimates their average height at thirteen to sixteen feet. We have seen that those of Baffin Bay are tenfold more considerable and imposing; but then, in that bay the temperature of the sea is below 32°; the glacier does not melt when it enters the water; it sinks to the bottom of the sea; and the portions detached from it are all of greater height than even the submerged part of the icebergs which drift to and fro in the bays and gulfs of Spitzbergen.

We may follow up this description with some observations by Lieutenant Bellot, the chivalrous young Frenchman who perished in one of the expeditions despatched in search of Sir John Franklin and his companions. He is speaking of the masses of ice his ship encountered soon after doubling Cape Farewell, the south point of Greenland, and he remarks, that as Baffin Bay narrows towards the south, the icebergs, first set in motion higher up the bay by the northern gales, necessarily tend to accumulate in the gorge thus formed, and so to impede and block up Davis Strait, even when the higher waters are quite free. It is only through a series of alternate movements of advance and recession that the bergs finally pass beyond the barrier, and float out into the Atlantic, to undergo a slow process of dissolution.

The mobility of the bergs, though necessary to navigation, forms at the same time its peculiar danger, since a vessel is often placed between the shore and the colossal masses driven forward by the wind, or between these and the solid ice which as yet has not broken up. It is useless to dwell upon the immense force possessed by masses which are frequently several square leagues in extent, and which, once in movement, cannot be stayed by any human resistance. A sailing-vessel finds herself placed in conditions all the more unfavourable, because the winds blow from the very direction which she is bound to take in order to open up a way through the floes. Now, if the gale is violent, it is perilous indeed to push forward in the midst of a labyrinth of bergs, which form so many floating rocks; if a calm prevails, a ship can move forward only by laborious hauling or towed by the boats. The application of the screw-propeller to steam-ships has given to them a great superiority, because they are not liable to any accident to paddle-wheels, exposed as such must be to collision with the floating ice. It is recorded that, on one occasion, a screw-steamer, near Upernavik, on the coast of Greenland, actually *charged* an iceberg, and drove right through it, as a railway-engine might crash through a fence or hurdle. Of course, the berg was of no great elevation; but its solid mass yielded to the immense force of the steam-ship, and split into large fragments.

In the convulsions caused by furious tempests, which are far from being so rare within the Arctic Circle as is popularly supposed, the shape of the bergs becomes very irregular, and the configuration of the ice-fields is constantly undergoing modification. Hence it often happens that the voyager sees before him an open basin of water of greater or less extent, from which he is separated only by a narrow strip of ice. In such a case he endeavours to effect an opening, either by driving his ship at full speed against the weakest part of the ice, or with the help of immense saws, twenty feet in length, which are worked with a rope and pulley placed at the top of a triangle formed of long poles; or, finally, by exploding a mine. When the ice is not very solid, the ship is forced into the opening, against the sides of which it acts like a wedge. It will sometimes occur, in the course of the operation, that the ice-fields, set in motion by the wind or the currents, close in together, after having treacherously separated for a moment, and the vessel is then subjected to a dangerous pressure. Unhappy the mariner who does not foresee or sufficiently note the warning signs of this accident, which is almost always accompanied by fatal consequences. The ice, which nothing can check, passing underneath the ship, capsizes it,—or, if it resists, crushes it.

We have alluded to the colossal bergs of Baffin Bay. These are thrown off from the northern glaciers, and particularly from the enormous ice-river named after Humboldt, which cumbers the declivities of the Greenland Alps, beyond the 79th parallel. It has been a frequent source of surprise to navigators that these mighty masses should float in a contrary direction to that of the ice-fields which descend with the Polar current towards the Atlantic. They reascend with such rapidity that they shatter the so-called "ice-foot," or belt of ice, still adhering to the shore. Captain Maury has collected numerous observations on this important subject, and he quotes the case of a ship which was being laboriously hauled against the current, when an enormous floating mountain coming up from the south steered against it, but fortunately did not come into collision with it, and forging ahead, very quickly disappeared. How is such an incident to be explained? By the existence of a submarine counter-current, acting on the lower extremity of the submerged portion of the berg, which, as we have stated, is always seven or eight times larger than the bulk *above* the surface of the waves.

Our whalers, in their hazardous expeditions, often derive assistance from these moving islands. They seek shelter under their lee when sudden storms arise; for the huge bergs are scarcely affected by the most violent gales. They find their shelter valuable also during certain operations of the fishery for which rest and quiet are necessary. Yet it is not absolutely exempt from danger. The seeming friend may prove to be a concealed foe. The iceberg may collapse, or be capsized; or formidable fragments, loosened from their sides or summits, may topple headlong and threaten to overwhelm the ship beneath: but as on these and other accidents we have already dwelt at length, we refrain from wearying our readers with a twice-told tale. The repetition in which, to some extent, we have indulged, was needful, in order to show the reader in what way the dissolution of the lower extremity of the glaciers is effected in the Arctic world.

In the neighbourhood of Cape Alexander, one of the headlands of Smith Strait, Dr. Hayes met with a glacier, of which he gives an interesting description in his narrative of an "Arctic Boat Journey," (1854):—

It was the first, protruding into the ocean, which he had had an opportunity of inspecting closely; and though small, compared with other similar formations, it had nevertheless all their principal characteristics. It presented to the sea a convex mural face, seventy feet in height and about two miles in length, its centre projecting into the water beyond the general line of the coast to the east and west of it. The surface rose abruptly to the height of about two hundred feet, and, sloping thence backward with a gentle inclination, seemed to be connected with an extensive *mer de glace* above. Several fissures or crevasses, apparently of great depth, struck vertically through its body, and extended far up into its interior; and others, more shallow, which

FORCING A PASSAGE THROUGH THE ICE.

seemed to have been formed by the streams of melted snow that poured in cataracts down into the sea. Dr. Hayes remarks that he was impressed by its viscous appearance; but we have shown that a certain amount of viscosity naturally appertains to glacier ice.

Parallel with its convex face ran a succession of indistinctly marked lines, which gave it the aspect of a semi-fluid mass moving downward upon an inclined surface; and this idea was confirmed by its appearance about the rocks on either side. Over these it seemed to have flowed; and, fitting accurately into all their inequalities, it gave the effect of a huge moving mass of partially solidified matter suddenly congealed.

Of still greater interest is the same adventurous explorer's description of the great Arctic *Mer de Glace* which lies inland from Rensselaer Bay, in about lat. 79° N., and long. 68° W.

Dr. Hayes and his party had set out on an expedition into the interior, and after passing through a really picturesque landscape, enriched with beds of moss and turf, patches of purple andromeda, and the trailing branches of the dwarf-willow, they emerged upon a broad plain or valley, in the heart of which reposed a frozen lake, about two miles in length by half a mile in width. They traversed its transparent surface. On either side of them rose rugged bluffs, that stretched off into long lines of hills, culminating in series in a broad-topped mountain-ridge, which, running away to right and left, was cut by a gap several miles wide that opened directly before them. Immediately in front was a low hill, around the base of which flowed on either side the branches of a stream whose course they had followed. Leaving the river-bed just above the lake, they climbed to the summit of this hillock; and there a sight burst upon them, grand and imposing beyond the power of words adequately to describe. From the rocky bed, only a few miles in advance, a sloping wall of pure whiteness rose to a broad level plain of ice, which, apparently without limits, stretched away toward the unknown east. It was the great *mer de glace* of the Arctic continent.

Here then was, in reality, the counterpart of the river-systems of other lands. From behind the granite hills the congealed drainings of the interior water-sheds, the atmospheric precipitations of ages, were moving in a mass, which, though solid, was plastic, moving down through every gap in the mountains, swallowing up the rocks, filling the valleys, submerging the hills: an onward, irresistible, crystal tide, swelling to the ocean. The surface was intersected by numerous vertical crevasses, some of considerable depth, which had drained off the melted snow.

It was midnight when the explorers approached this colossal reservoir. The sun was several degrees beneath the horizon, and afforded a faint twilight-gleam. Stars of the second magnitude were dimly perceptible in the cold, steel-blue Arctic heavens. When they were within about half a mile of the icy wall, a brilliant meteor fell before them, and, by its reflection upon the glassy surface beneath, greatly increased the magical effect of the scene; while loud reports, like distant thunder or the roll of artillery, broke at intervals from the depths of the frozen sea.

On closer inspection it was found that the face of the glacier ascended at an angle of from 30° to 35°. At its base lay a high bank of snow, and the wanderers clambered up it about sixty feet; but beyond this their efforts were defied by the exceeding smoothness of the ice. The mountains, which stood on either hand like giant-warders, were overlapped, and to some extent submerged, by the glacier. From the face of the huge ice-river innumerable little rivulets ran down the channels their action had gradually excavated, or gurgled from beneath the ice; forming, on the level lands below, a sort of marsh, not twenty yards from the icy wall. Here, in strange contrast, bloomed beds of verdurous moss; and in these, tufts of dwarf-willows were wreathing their tiny arms and rootlets about the feebler flower-growths; and there, clustered together, crouching among the grass, and sheltered by the leaves, and feeding on the bed of lichens, flourished a tiny, white-blossomed draba and a white chickweed. Dotting the few feet of green around might be seen the yellow flowers of the more hardy poppy, the purple potentilla, and saxifrages yellow, purple, and white.

The great glacier of Sermiatsialik is one of the arms, or outlets, of this immense reservoir of ice. It occupies the bed of a valley, varying from three and a half to five miles in width, and attaining at certain points a depth of upwards of three hundred and seventy feet. This valley opens upon the fiord of Sermiatsialik, which is separated from that of Julianshaab by the range of mountains culminating in the peak of Redkammen.

We owe to Dr. Hayes a lively description of the Sermiatsialik glacier, which he thinks must at some places be more than seven hundred and fifty feet in depth, overflowing the borders of the valley like a swollen torrent. For upwards of four leagues, the icebergs which throng the fiord, or gulf, are those of the glacier itself, and terminating in a wedge-like outline, disappear in the vast sea of ice expanding to right and left above the loftiest summits, and drawing irresistibly the eye to its rippled surface,—boundless, apparently, like that of ocean. As the voyager sails up the gulf, he gradually loses sight of the frozen slope, and then of the white line of the *mer de glace*: he finds himself in front of an immense cliff, from one hundred to two hundred feet in height, diaphanous as the purest crystals, and reflecting all the hues of heaven.

One almost shudders as one approaches this vast domain of Winter. Collecting in copious streams, the ice and snow melted on the surface of the glacier pour over its brink, forming floating clouds of spray, irradiated by rainbow colours. The din of these cascades fills the air. At intervals, the loud reports of the internal convulsions of the glacier are repeated by every echo.

The cliff is entirely vertical; but its face, far from being smooth, is broken up into an infinite variety of forms: into unfathomable cavernous hollows, symmetrical spires, ogives, pinnacles, and deep fissures, where the eye plunges into a transparent blue, which changes every second its fleeting, opaline tints; tints so soft, and yet so vivid, that they defy the skill of the artist to reproduce them. The lustre of the "dark eye of woman" is not more difficult to seize. A deep dark green, less delicate but not less splendid, colours all the recesses where the ice overhangs the waters. In the sunlight one sees the surface of these huge crystals shining with the whiteness of the purest snow; except, indeed, where recent fractures have taken place. They suggest to the mind the idea of the gleams and reflections of a piece of satin; the undulatory lustre and shifting sparkle being produced by the different angles under which the light is reflected.

But let us suppose that we have landed; with much difficulty have ascended the cliffs; and have clambered up the glacier to its very summit. The scene before us, how shall we convey to the mind of the reader?

Imagine, if you can, the rapids of the Upper Niagara congealed even to their lowest depths; imagine the falls, and the broad river, and the great Lake Erie all frozen into solid ice; with

THE GLACIER OF SERMIATSIALIK, GREENLAND.

bergs above the cataract towering as high as the lower banks: suppose that you, the spectator, having taken your stand upon the rapids, with the Erie so near that you can see its crystallized surface, and you will have a picture, on a reduced scale, of the sea of ice now spreading far before us. The rapids will represent the glacier; the Great Fall the cliff which it projects into the sea (only that the celebrated "horse-shoe" is here turned outwards); the river which broadens into the Ontario will be the fiord; and the Ontario, that dark grim ocean into which the gigantic bergs detached from the mighty ice-cascade are slowly making their way!

We must indicate, however, one remarkable dissimilarity, for which our previous observations on the nature of glaciers will have prepared the reader. From one bank to the other, the surface of a river is always *horizontal*, but that of a glacier is slightly *convex*.

Through the narrow glen, or ravine, formed by this curvature of the glacier, a kind of lateral trough or gully, bounded by the escarpment of the soil, we reach the sea. The descent is not without its dangers, for at every point crevasses open, separated by slippery projections. These deep gashes, at some points, are only a few yards apart; and they incessantly cross each other, and run into one another, so as to form a perfect labyrinth, in the windings of which the adventurous traveller is apt to feel bewildered.

The border of the glacier once crossed, the way becomes less difficult; for a mile and a half the level is almost perfect, and the ice but little broken up. The frozen desert, however, impresses us with an almost solemn feeling, and there is something terrible in the desolation of such a Sahara of snow!

Moreover, the traveller is irresistibly affected by the continual roar or growling of the enormous mass, which seems to stir and shake under our very feet. He would not be surprised if a vast chasm suddenly yawned before him! These harsh deep voices of the glacier, however, are not the only sounds we hear. On every side rises the murmur of brooks which trace their furrows across the crystalline plain. Some of these gradually converge, and, uniting, form a considerable torrent, which leaps with a clang from icy crag to icy ledge, until it is lost in a crevasse, or precipitated over the frozen cliff into the waters of the fiord. The *solitude* of the scene is complete, but not the *silence*. The air is as full of "noises" as ever was Prospero's isle.

Such are the principal features of the glacier of Sermiatsialik.

About ninety miles north-east of Rensselaer Bay lies the great Humboldt Glacier, which seems to serve as a connecting-link between the Old World and the New.

It lies between the 79th and 80th parallels north, and between the 64th and 65th meridians west, skirting the shore of Peabody Bay, which is a bold indentation of the east coast of Kane Sea.

It was discovered in Dr. Kane's expedition, and is probably one of the grandest spectacles in the Arctic world. Dr. Kane acknowledges himself unable to do justice to its magnificent aspect. He can speak only of its "long, ever-shining line of cliff diminished to a well-pointed wedge in the perspective;" of its "face of glistening ice, sweeping in a long curve from the low interior, the facets in front intensely illuminated by the sun."

This line of cliff rises, like a solid wall of glass, three hundred feet above the water-level, with an unknown, unfathomable depth below it; and its curved face, *sixty miles in length*, disappears into unknown space at not more than a single day's railroad-travel from the Pole. The interior with which it communicates, and from which it issues, is an unexplored *mer de glace*, an ice-ocean, of apparently boundless dimensions.

Such is the "mighty crystal bridge" which connects the two continents of America and Greenland. We say, continents; for Greenland, as Dr. Kane remarks, however insulated it may ultimately prove to be, is in mass strictly continental. Its least possible axis, measured from Cape Farewell to the line of the Humboldt Glacier, in the neighbourhood of the 80th parallel, gives a length of upwards of twelve hundred miles,—not materially less than that of Australia from its northern to its southern cape.

Imagine the centre of such a continent, says Dr. Kane, occupied through nearly its whole extent by a deep, unbroken sea of ice, that gathers perennial increase from the water-shed of vast snow-covered mountains and all the precipitations of the atmosphere upon its own surface. Imagine this, moving onward like a great glacial river, seeking outlets at every fiord and valley, rolling icy cataracts into the Atlantic and Greenland seas; and, having at last reached the northern limit of the land that has borne it up, pouring out a mighty frozen torrent into unknown Arctic space.

"It is thus," remarks Dr. Kane, "and only thus, that we must form a just conception of a phenomenon like this great glacier. I had looked in my own mind for such an appearance, should I ever be fortunate enough to reach the northern coast of Greenland. But now that it was before me, I could hardly realize it. I had recognized, in my quiet library at home, the beautiful analogies which Forbes and Studer have developed between the glacier and the river; but I could not comprehend at first this complete substitution of ice for water.

"It was slowly the conviction dawned on me that I was looking upon the counterpart of the great river-system of Arctic Asia and America. Yet here were no water-feeders from the south. Every particle of moisture had its origin within the Polar Circle, and had been converted into ice. There were no vast alluvions, no forest or animal traces borne down by liquid torrents. Here was a plastic, moving, semi-solid mass, obliterating life, swallowing rocks and islands, and ploughing its way with irresistible march through the crust of an investing sea."

When, at a later period, Dr. Kane made a closer examination of this great natural wonder, he found that previously he had not realized the full grandeur of the spectacle. He noted that the trend of the glacier was a few degrees to the west of north; and he remarks, as the peculiarity of its aspect, that it did not indicate *repose*, but activity, energy, movement.

Its surface seemed to follow that of the basis-country over which it flowed. It was undulating on and about the horizon, but as it descended towards the sea it represented a broken plain with a general inclination of some nine degrees, still diminishing toward the foreground. Crevasses, which in the distance seemed like mere wrinkles, expanded as they came nearer, and were intersected almost at right angles by long continuous lines of fracture parallel with the face of the glacier.

These lines, too, scarcely perceptible in the far distance, widened as they approached the sea until they formed a gigantic stairway. It seemed as though the ice had lost its support below, and that the mass was let down from above in a series of steps; and such an action is the necessary result of the heat thrown out by the soil, the excessive surface-drainage, and the constant abrasion of the sea.

The indication of a great propelling agency seemed to be just commencing at the time that Dr. Kane visited the great glacier. The split-off lines of ice were evidently in motion, pressed on by those behind, but still broadening their fissures, as if the impelling action grew more and more energetic nearer the water, till at last they floated away in the form of icebergs. Long files of these detached masses might be seen, like the ranks of a stately armada, slowly sailing out into the remote sea, their separation marked by dark parallel shadows; broad and spacious avenues near the eye, but narrowed in the perspective to mere furrows. A more impressive illustration of the forces of nature it would be difficult to conceive.

Dr. Kane's view of the formation of icebergs differs considerably from that which most physicists entertain.

He does not believe that the berg falls into the sea, broken by its weight from the parent glacier; he is of opinion that it *rises from* the sea. The process is at once gradual and comparatively quiet. "The idea of icebergs being discharged, so universal among systematic writers, seems to me at variance with the regulated and progressive actions of Nature. Developed by such a process, the thousands of bergs which throng the Polar seas should keep the air and water in perpetual commotion, one fearful succession of explosive detonations and propagated waves. But it is only the lesser masses falling into deep waters which could justify the popular opinion. The enormous masses of the great glacier are propelled, step by step and year by year, until, reaching water capable of supporting them, they are floated off to be lost in the temperatures of other regions."

The Humboldt Glacier did not differ in structure from the Alpine and Norwegian ice-growths; and its face presented nearly all the characteristic features of the latter. The *overflow*, or viscous overlapping of the surface, was very strongly marked. "When close to the island rocks," says Kane, "and looking out upon the upper table of the glacier, I was struck with the homely analogy of the batter-cake spreading itself out under the ladle of the housewife, the upper surface less affected by friction, and rolling forward in consequence."

The crevasses bore the marks of direct fracture, as well as of the more gradual action of surface-drainage. The extensive water-shed between their converging planes gave to the icy surface most of the hydrographic features of a river-system. The ice-born rivers which divided them were margined occasionally with spires of discoloured ice, and generally lost themselves in the central areas of the glacier before reaching its foreground. Occasionally, too, the face of the glacier was cut by vertical lines, which, as in the Alpine examples, were evidently outlets for the surface drainage.

The height of this ice-wall at the nearest point was about three hundred feet, measured from the water's edge; and the unbroken right line of its diminishing perspective showed that this might be regarded as its constant measurement. It seemed, in fact, a great icy table-land, abutting with a clean precipice against the sea. This, indeed, is the great characteristic of all those Arctic glaciers which issue from central reservoirs or *mers de glace* upon the fiords or bays, and is strikingly in contrast with the dependent or hanging glacier of the ravines, where every line and furrow and chasm seem to indicate the movement of descent and the mechanical disturbances which have impeded and delayed it.

Dr. Kane named this monster glacier after Alexander Von Humboldt, to whose labours

Physical Science is so largely indebted; and the cape which flanks it on the Greenland coast after the distinguished naturalist, whom the world has so recently lost, Professor Agassiz.

The point at which the Humboldt Glacier enters the "Land of Washington" affords even at a distance very clear indications of its plastic or semi-solid character. The observer finds it impossible to resist the impression of fluidity conveyed by its peculiar markings. Dr. Kane very appropriately named it Cape Forbes, in honour of the illustrious son of Scotia who contributed so largely to our true knowledge of the structure and mode of progression of glaciers.

As the surface of the glacier, adds its discoverer, receded to the south, its face seemed broken with piles of earth and rock-stained rubbish, until far back in the interior it was concealed from view by the slope of a hill. But even beyond this point its continued extension was shown by the white glare or ice-blink in the sky above.

Its outline to the northward could not be so easily traced, on account of the enormous discharges at its base. The talus of its descent from the interior, looking far off to the east, ranged from 7° to 15°; so interrupted by the crevasses, however, as only in the distance to produce the effect of an inclined plane. A few black protuberances rose above the glittering surface of the snow, like islands in a foamy sea.

It could be seen, from the general inequalities of its surface, how well the huge mass adapted itself to the inequalities of the basis-country beneath. The same modifications of hill and dale were discernible as upon land. Thus grand and various in its imposing aspect, it stretches to the north until it touches the new Land of Washington, cementing together by an apparently indissoluble tie the Greenland of the Norse Vikings and the America of the Anglo-Saxon colonists.

CHAPTER V.

THE ARCTIC LANDS—FAUNA—FLORA—GREENLAND—ICELAND—NOVAIA ZEMLAIA—SIBERIA.

WE have already pointed out that in the northernmost regions of the Arctic lands the year is divided into one prolonged and bitterly cold night of several months' duration, and one glorious summer's day extending over nine or ten weeks, which brings the scanty vegetation to a sudden maturity. We have indicated that even within the limits of perpetual snow the life of Nature is not altogether crushed out; and in support of this statement we may refer to the "red snow" which figures so often in the pages of our Arctic voyagers, though its true character was not at first apprehended.

This so-called "red snow" was found by Sir John Ross, in his first Arctic expedition in 1808, on a range of cliffs rising about 800 feet above the sea-level, and extending eight miles in length (lat. 75° N.). It was also discovered by Sir W. E. Parry in his overland expedition in 1827. The snow was tinged to the depth of several inches. Moreover, if the surface of the snow-plain, though previously of its usual spotless purity, was crushed by the pressure of the sledges and of the footsteps of the party, blood-like stains instantly arose; the impressions being sometimes of an orange hue, and sometimes more like a pale salmon tint.

It has been ascertained that this singular variation of colour is due to an immense aggregation of minute plants of the species called *Protococcus nivalis;* the generic name alluding to the extreme primitiveness of its organization, and the specific to the peculiar nature of its habitat. If we place a small quantity of red snow on a piece of white paper, and allow it to melt and evaporate, there will be left a residuum of granules sufficient to communicate a faint crimson tint to the paper. Examine these granules under a microscope, and they will prove to be spherical purple cells of almost inappreciable size, not more than the three-thousandth to one-thousandth part of an inch in diameter. Look more closely, and you will see that each cell has an opening, surrounded by indented or serrated lines, the smallest diameter of which measures only the five-thousandth part of an inch. When perfect, the plant, as Dr. Macmillan observes, bears a resemblance to a red-currant berry; as it decays, the red colouring matter fades into a deep orange, which is finally resolved into a brownish hue. The thickness of the wall of the cell is estimated at the twenty-thousandth part of an inch, and three hundred to four hundred of these cells might be grouped together in a smaller space than a shilling would cover. Yet each cell is a distinct individual plant; perfectly independent of others with which it may be massed; fully capable of performing for and by itself all the functions of growth and reproduction; possessing "a containing membrane which absorbs liquids and gases from the surrounding matrix or elements, a contained fluid of peculiar character formed out of these materials, and a number of excessively minute granules equivalent to spores, or, as some would say, to cellular buds, which are to become the germs of new plants." Dr. Macmillan adds: "That one and the same primitive cell should thus minister equally to absorption, nutrition, and reproduction, is an extraordinary illustration of the fact that the smallest and simplest organized object is in itself, and, for the part it was created to perform in the operations of nature, as admirably adapted as the largest and most complicated."

PROTOCOCCUS NIVALIS.

The first vegetable forms to make their appearance at the limits of the snow-line, whether in high latitudes or on mountain-summits, are lichens; which flourish on rocks, or stones, or trees, or wherever they can obtain sufficient moisture to support existence. Upwards of two thousand four hundred species are known. The same kinds prevail throughout the Arctic Regions, and the species common to both the Eastern and Western Hemispheres are very numerous. They lend the beauty of colour to many an Arctic scene which would otherwise be inexpressibly dreary; the most rugged rock acquiring a certain air of picturesqueness through their luxuriant display. Their forms are wonderfully varied; so that they present to the student of Nature an almost inexhaustible field of inquiry. In their most rudimentary aspects they seem to consist of nothing more than a collection of powdery granules, so minute that the figure of each is scarcely distinguishable, and so dry and so deficient in organization that we cannot but wonder how they live and maintain life. Now they are seen like ink-spots on the trunks of fallen trees; now they are freely sprinkled in white dust over rocks and withered tufts of moss, others appear in gray filmy patches; others again like knots or rosettes of various tints; and some are pulpy and gelatinous, like aërial sea-weeds which the receding tide leaves bare and naked on inland rocks. A greater complexity of structure, however, is visible in the higher order of lichens,—and we find them either tufted and shrubby, like miniature trees; or in clustering cups, which, Hebe-like, present their "dewy offerings to the sun."

In the Polar World, and its regions of eternal winter, where snow and ice, and dark drear waters, huge glacier and colossal berg, combine to form an awful and impressive picture, the traveller is thankful for the abundance of these humble and primitive forms, which communicate the freshness and variety of life to the otherwise painful and death-like uniformity of the frost-bound Nature. It is true that here,

"Above, around, below,
On mountain or in glen,
Nor tree, nor shrub, nor plant, nor flower."

may be found in the lands beyond the line of perpetual snow; it is true that

"All is rocks at random thrown,
Black waves, bare crags, and banks of stone;
As if were here denied
The summer's sun, the spring's sweet dew,
That clothe with many a varied hue
The bleakest mountain-side;"

but vegetation is not absolutely wanting, and the lichens are so largely developed and so widely distributed as to impart quite a peculiar and distinctive character to the scenery.

A lichen which is discovered in almost every zone of altitude and latitude, which ranges from the wild shores of Melville Island in the Arctic to those of Deception Island in the Antarctic circle,—which blooms on the crests of the Himalayas, on the lofty peak of Chimborazo, and was found by Agassiz near the top of Mont Blanc,—is the *Lecidea geographica*, a beautiful bright-green lichen, whose clusters assume almost a kaleidoscopic appearance.

A lichen of great importance in the Arctic world is the well-known *Cladonia rangiferina*, or reindeer moss, which forms the staple food of that animal during the long Arctic winter. In the vast tundras, or steppes, of Lapland it flourishes in the greatest profusion, completely covering the ground with its snowy tufts, which look like the silvery sprays of some magic plant. According to Linnæus, it thrives more luxuriantly than any other plant in the pine-forests of Lapland, the surface of the soil being carpeted with it for many miles in extent; and if the forests are accidentally burned to the ground, it quickly reappears, and grows with all its original vigour. These plains, which seem to the traveller smitten with the curse of desolation, the Laplander regards as fertile pastures; and here vast herds of reindeer roam at will, thriving where the horse, the elephant, and even the camel would perish. This useful animal is dependent almost entirely on a lichen for support. What a deep interest is thus attached to it! That vast numbers of families, living in pastoral simplicity in the cheerless and inhospitable Polar Regions, should depend for their subsistence upon the uncultured and abundant supply of a plant so low in the scale of organization as this, is, says Dr. Macmillan, a striking proof of the great importance of even the smallest and meanest objects in nature.

When the ground is crusted with a hard and frozen snow, which prevents it from obtaining its usual food, the reindeer turns to another lichen, called rock-hair (*Alectoria jubata*), that grows in long bearded tufts on almost every tree. In winters of extreme rigour, the Laplanders cut down whole forests of the largest trees, that their herds may browse freely on the tufts which clothe the higher branches. Hence it has been justly said that "the vast dreary pine-forests of Lapland possess a character which is peculiarly their own, and are perhaps more singular in the eyes of the traveller than any other feature in the landscapes of that remote and desolate region. This character they owe to the immense number of lichens with which they abound. The ground, instead of grass, is carpeted with dense tufts of the reindeer moss, white as a shower of new fallen snow; while the trunks and branches of the trees are swollen far beyond their natural dimensions with huge, dusky, funereal branches of the rock-hair, hanging down in masses, exhaling a damp earthy smell, like an old cellar, or stretching from tree to tree in long festoons, waving with every breath of wind, and creating a perpetual melancholy sound."

In regions furthest north are found various species of lichens belonging to the genera *Gyrophora* and *Umbilicaria*, and known in the records of Arctic travel as rock tripe, or *tripe de roche;* a name given to them in consequence of their blistered thallus, which bears a faint resemblance to the animal substance so called. They afford a coarse kind of food, and proved of the greatest service to the expeditions under Sir John Franklin; though their nutritious properties are not considerable, and, such as they are, are unfortunately impaired by the presence of a bitter principle which is apt to induce diarrhœa. In Franklin and Richardson's terrible overland journey from the Coppermine River to Fort Enterprise it was almost the sole support, at one time, of the heroic little company. Dr. Richardson says they gathered four species of

Gyrophora,* and used them all as articles of food; "but not having the means of extracting the bitter principle from them, they proved nauseous to all, and noxious to several of the party, producing severe bowel complaints." Franklin on one occasion remarks: "This was the sixth day since we had enjoyed a good meal; the *tripe de roche*, even when we got enough, only serving to allay the pangs of hunger for a short time." Again, we read: "The want of *tripe de roche* caused us to go supperless to bed."

Dr. Hayes, in the course of his "Arctic Boat Journey," was compelled to have recourse to the same unsatisfactory fare. The rock-lichen, or stone-moss, as he calls it, he describes as about an inch in diameter at its maximum growth, and of the thickness of a wafer. It is black externally, but when broken the interior appears white. When boiled it makes a glutinous fluid, which is slightly nutritious.

"Although in some places it grows very abundantly," writes Dr. Hayes, "yet in one locality it, like the game, was scarce. Most of the rocks had none upon them; and there were very few from which we could collect as much as a quart. The difficulty of gathering it was much augmented by its crispness, and the firmness of its attachment.

"For this plant, poor though it was, we were compelled to dig. The rocks in every case were to be cleared from snow, and often our pains went unrewarded. The first time this food was tried it seemed to answer well,—it at least filled the stomach, and thus kept off the horrid sensation of hunger until we got to sleep; but it was found to produce afterwards a painful diarrhœa. Besides this unpleasant effect, fragments of gravel, which were mixed with the moss, tried our teeth. We picked the plants from the rock with our knives, or a piece of hoop-iron; and we could not avoid breaking off some particles of the stone."

These lichens are black and leather-like, studded with small black points like "coiled wire buttons," and attached either by an umbilical root or by short and tenacious fibres to the rocks. Some of them may be compared to a piece of shagreen, while others resemble a fragment of burned skin. They are met with in cold bleak localities, on Alpine heights of granite or micaceous schist, in almost all parts of the world,—on the Scottish mountains, on the Andes, on the Himalayas; but it is in the Polar World that they most abound, spreading over the surface of every rock a sombre Plutonian vegetation, that seems to have been scathed by fire and flame, until all its beauty and richness were shrivelled up.

Some of the lichens in the less remote latitudes—as, for instance, in Sweden—are far superior in usefulness to any of those we have hitherto described. The Swedish peasant finds in them his pharmacy, his dyeing materials, his food. With the various lichens that grow upon the trees and rocks, says Frederika Bremer, he cures the virulent diseases which sometimes afflict him, dyes the articles of clothing which he wears, and poisons the noxious and dangerous animals which annoy him. The juniper and cranberry give him their berries, which he brews into drink; he makes a conserve of them, and mixes their juices with his dry salt-meat, and is healthful and cheerful with these and with his labour, of which he makes a pleasure.

The only lichen which has retained its place in modern pharmacy is the well-known "Iceland moss." It is still employed as a tonic and febrifuge in ague; but more largely, when added to soups and chocolate, as an article of diet for the feeble and consumptive. In Iceland the

* So called from its circular form, and because the surface of the leaf is marked with curved lines.

Cetraria Islandica is highly valued by the inhabitants. What barley, rye, and oats are to the Indo-Caucasian races of Asia and Western Europe; the olive, the fig, and the grape to the inhabitants of the Mediterranean basin; rice to the Hindu; the tea-plant to the native of the Flowery Land; and the date-palm to the Arab,—is Iceland moss to the Icelander, the Lapp, and the Eskimo.

It is found on some of the loftiest peaks of the Scottish Highlands; but in Iceland it overspreads the whole country, flourishing more abundantly and attaining to a larger growth on the volcanic soil of the western coast than elsewhere. It is collected triennially, for it requires three years to reach maturity, after the spots where it thrives have been cleared. We are told that the meal obtained from it, when mixed with wheat-flour, produces a greater quantity, though perhaps a less nutritious quality, of bread than can be manufactured from wheat-flour alone. The great objection to it is its bitterness, arising from its peculiar astringent principle, cetraria. However, the Lapps and Icelanders remove this disagreeable pungency by a simple process. They chop the lichen to pieces, and macerate it for several days in water mixed with salt of tartar or quicklime, which it absorbs very readily; next they dry it, and pulverize it; then, mixed with the flour of the common knot-grass, it is made into a cake, or boiled, and eaten with reindeer's milk.

Mosses are abundant in the Arctic Regions, increasing in number and beauty as we approach the Pole, and covering the desert land with a thin veil of verdure, which refreshes the eye and gladdens the heart of the traveller. On the hills of Lapland and Greenland, they are extensively distributed; and the landscape owes most of its interest to the charming contrasts they afford. Of all the genera, perhaps the bog-mosses, *Sphagna*, are the most luxuriant; but at the same time they are the least attractive, and the plains which they cover are even drearier than the naked rock. In Melville Island these mosses form upwards of a fourth part of the whole flora. Much finer to the sight is the common hair-moss (*Polytrichum commune*), which extends over the levels of Lapland, and is used by the Lapps, when they are bound on long journeys, for a temporary couch. We may mention also the fork-moss (*Dicranum*), which the Eskimos twist into wicks for their rude lamps.

We have not space to dwell upon the grasses and fungi, though these are numerous, and some of them interesting. The *cochlearia*, or scurvy-grass, has often proved of great utility to Arctic explorers; and Dr. Kane on more than one occasion availed himself of its medicinal properties. Fungi extend almost to the very limits of Arctic vegetation. The Greenlanders and Lapps make use of them for tinder, or as styptics for stopping the flow of blood, and allaying pain. In Siberia they abound. Frequently, in the high latitudes, they take the form of "snow mould," and are found growing on the barren and ungenial snow. These species are warmed into life only when the sun has grown sufficient to melt the superficial snow-crust, without producing a general thaw, and then they spread far and wide in glittering wool-like patches, dotted with specks of red or green. When the snow melts, they overspread the grass beneath like a film of cobweb, and in a day or two disappear.

In Siberia grows the fly-agaric (*Agaricus muscarius*), from which the inhabitants obtain an intoxicating liquor of peculiarly dangerous character. It has a tall white stem, surmounted by a dome of rich orange scarlet, studded with white scaly tubercles, and in some parts of Kamtschatka and the northern districts of Siberia is so abundant that the ground sparkles and shines as if covered with a scarlet carpet. The natives collect it during the hot summer months, and dry it. Steeped in the juice of the whortleberry, it forms a powerful intoxicating wine; or rolled up like a bolus, and swallowed without chewing, it produces much the same effect as opium. On some, however, it acts as an excitant, and induces active muscular exertion. A talkative person, under its influence, cannot keep silence or secrets; one fond of music, sings incessantly; and if a person who has partaken of it wishes to step over a straw or small stick, he takes a stride or jump sufficient to clear the trunk of a tree!

The Koriaks and Kamtschatkans personify this fungus, under the name of *Mocho Moro*, as one of their *penates*, or household gods; and if they are impelled by its effects to commit any dreadful crime, they pretend they act only in obedience to commands which may not be disputed. To qualify themselves for murder or suicide, they drink additional doses of "this intoxicating product of decay and corruption."

During Captain Penny's voyage in search of Sir John Franklin, he picked up two pieces of floating drift-wood, far beyond the usual limit of Eskimo occupation, which, from their peculiar appearance, excited a lively curiosity. The one was found in Robert Bay, off Hamilton Island, lat. 76° 2′ north, and long. 76° west,—that is, in the route which Franklin's ships, it is supposed, had followed,—and was plainly a fragment of wrought elm plank, which had been part of a ship's timbers. It exhibited three kinds of surface,—one that had been planed and pitched, one roughly sawn, and the third split with an axe. The second piece of drift-wood was picked up on the north side of Cornwallis Island, in lat. 75° 36′ north, and long. 96° west. It was a branch of white spruce, much bleached in some places, and in others charred and blackened as if it had been used for fuel.

On both fragments traces of microscopic vegetation were discovered; and as it was thought they might, if carefully examined, afford some clue to the fate of Franklin's expedition, they were submitted to Mr. Berkeley, a well-known naturalist. In the report which he addressed to the Admiralty, he stated that the vegetation in both cases resembled the dark olive mottled patches with which wooden structures in this country, if exposed to atmospheric influences, are speedily covered. The bleached cells and fibres of the fragment of elm were filled up with slender fungoid forms, *myceliæ;* while on its different surfaces appeared several dark-coloured specks, belonging to the genus *Phoma*. As it was not probable that plants so minute could have retained, through the terrible severity of an Arctic winter, their delicate naked spores in the perfect condition in which they were found, Mr. Berkeley concluded that they must have been developed through that same summer; while from three to four years, in those high latitudes and amid the rigour of stormy ice-covered seas, would suffice to produce the bleached appearance of the wood. Hence he inferred that the plank had not been long exposed.

On the other fragment of drift-wood he discovered some deeply-embedded minute black fungoid forms, called *Sporidesmium lepraria*. Unlike the *phomas*, which are very ephemeral, these plants possess the longevity of the lichens, and the same patches last for years unchanged on the same pieces of wood, while their traces are discernible for a still longer period. From their condition, Mr. Berkeley inferred that the fungi on the drifted wood had not been recently developed, but that, on the contrary, they were the remains of the species which existed on the drift-wood when used for fuel by the unfortunate crews of Franklin's ships, the *Erebus* and the *Terror*.

There can be no doubt whatever, as Dr. Macmillan remarks, considering the circumstances in which they were discovered, and the remarkable appearances they presented—there can be no reasonable doubt that both fragments of drift-wood belonged to, or were connected with, the lost ships; and the curious information regarding the course they pursued at a certain time, furnished by witnesses so extraordinary and unlikely as a few tiny dark specks of cryptogamic vegetation on floating drift-wood, was confirmed, in a wonderful manner, by the after-discovery of the first authentic account ever obtained of the sad and pathetic history of Franklin's expedition.

The reader will not expect to find the tundras of Northern Asia or the shores of the Polar Sea rich in bud and bloom, yet even these dreary wastes are not absolutely without floral decoration. Selinum and cerathium, as well as the poppy and sorrel, andromeda, and several species of heath, are mentioned by Dr. Kane as blooming in the neighbourhood of Smith Strait. On the south coast of the Polar Sea Dr. Richardson found a considerable variety of vegetation. We noticed, he says, about one hundred and seventy phænogamous or flowering plants; being one-fifth of the number of species which exist fifteen degrees of latitude further to the southward. He adds:—The grasses, bents, and rushes constitute only one-fifth of the number of species on the coast, but the two former tribes actually cover more ground than all the rest of the vegetation. The *cruciferæ*, or cross-like tribe, afford one-seventh of the species, and the compound flowers are nearly as numerous. The *shrubby plants* that reach the sea-coast are the common juniper, two species of willow, the dwarf-birch, the common alder, the hippophaë, a gooseberry, the red bear berry (*arbutus uva ursi*), the Labrador tea-plant, the Lapland rose, the bog-whortleberry, and the crowberry. The kidney-leaved oxyria grows in great abundance there, and occasionally furnished us with an agreeable addition to our meals, as it resembles the garden-sorrel in flavour, but is more juicy and tender. It is eaten by the natives, and must, as well as many of the cress-like plants, prove an excellent corrective of the gross, oily, rancid, and frequently putrid meat on which they subsist. The small balls of the Alpine bistort, and the long, succulent, and sweet roots of many of the astragaleæ, which grow on the sandy shores, are eatable; but it does not seem that the Eskimos are acquainted with their use. A few clumps of white spruce-fir, with some straggling black spruces and canoe-birches, grow at the distance of twenty or thirty miles from the sea, in sheltered situations on the banks of rivers.

It has been pointed out that the principal characteristic of the vegetation of the Arctic Regions is the predominance of perennial and cryptogamous plants; but further southward, where night begins to alternate with day, or in what may be called the sub-arctic zone, a difference of species appears which greatly enhances the beauty of the landscape. A rich and vividly-coloured flora adorns these latitudes in Europe as well as in Asia during their brief but ardent summer, with its intense radiance and intense warmth,—consisting of potentillas, gentians, starry chick-weeds, spreading saxifrages and sedums, spiræas, drabas, artemisias, and the like. The power of the sun is so great, and the consequent rapidity of growth so extraordinary, that these plants spring up, and blossom, and germinate, and perish in six weeks. In a lower latitude many ligneous plants are found,—as berry-bearing shrubs, the glaucous kalmia, the trailing azalea, the full-blossomed rhododendron. The Siberian flora differs from the European in the same latitudes by the inclusion of the North American genera, phlox, mitella, and claytonia, and by the luxuriance of its asters, spiræas, milk-vetches, and the saline plants goosefoot and saltwort.

In Novaia Zemlaia and other northern regions the vegetation is so stunted that it barely covers the ground, but a much greater variety of minute plants of considerable beauty are aggregated there in a limited space than in the Alpine climes of Europe where the same genera occur. This is due to the feebleness of the vegetation; for in the Swiss Alps the same plant frequently usurps a large area, and drives out every other,—as the dark blue gentian, the violet-tinted pansy, and the yellow and pink stone-crops. But in the far north, where vitality is weak and the seeds do not ripen, thirty different species, it has been observed, may be seen "crowded together in a brilliant mass," no one being powerful enough to overcome its companions. In these frozen climates plants may be said to live between the air and the earth, for they scarcely raise their heads above the soil, and their roots, unable to penetrate it, creep along the surface. All the woody plants—as the betula nava, the reticulated willow, andromeda tetragona, with a few bacciferous shrubs—trail upon the ground, and never rise more than an inch or two above it. The *Salix lanata*, the giant of the Arctic forests, is about five inches in height; while its stem, ten or twelve feet long, lies hidden among the moss, and owes shelter, almost life, to its humble neighbour.

From Novaia Zemlaia we pass to Spitzbergen, whose flora contains about ninety-three species of flowering or phænogamous plants, which, like those already mentioned, generally grow in tufts or patches, as if for the sake of mutual protection. The delicate mosses which clothe the moist lowlands, and the hardy lichens which incrust the rocks up to the remotest limits of vegetation, are very numerous. Some of the Spitzbergen plants are found on the Alps, at elevations varying from 9000 to 10,000 feet above the sea-level; such as the *Arenaria biflora*, the *Cerastium alpinum*, and the *Ranunculus glacialis*. The only esculent plant is the *Cochlearia fenestrata*, which here loses its bitter principles, so much complained of by our Arctic explorers, and may be eaten as a salad. Iceland moss and several grasses afford sustenance for the reindeer.

A very different description is given of Kamtschatka, to which we are once more brought in the course of our rapid survey. Its climate is much more temperate and uniform than that of Siberia, and as the air is humid, the herbaceous vegetation is extraordinarily luxuriant. Not only along the banks of the rivers and lakes, but in the avenues and copses of the woodlands, the grass attains a height of fully twelve feet, while the size of some of the compositæ and umbelliferæ is really colossal. For example, the *Heraclium dulce* and the *Senecio cannabifolius* frequently grow so tall as to overtop a rider upon horseback. The pasturage is so rich that the grass generally yields three crops every summer. A species of lily, the dark purple *Fritallaria sarrana*, is very abundant, and the inhabitants use its tubers instead of bread and meal. If the fruits of the bread-fruit tree are pre-eminent among all others, as affording man a perfect substitute for bread, the roots of the sarrana, which are very similar in taste, rank perhaps immediately after them. The collection of these tubers in the meadows is an important summer occupation of the women, and one which is rather troublesome, as the plant never grows gregariously, so that each root has to be dug out separately with a knife. Fortunately the work of gathering the tubers is much lightened by the activity of the Siberian field-vole, which excavates an ample burrow, and stores it for winter provision with a large supply of roots, chiefly those of the sarrana.

To sum up:—

What may be called the Arctic climate extends over nearly the whole of Danish America, the newly-acquired possessions of the United States, the original Hudson Bay Territory, and Labrador, down to that unimportant watershed which separates from the tributaries of Hudson Bay the three great basins of the St. Lawrence, the five great lakes, and the Mississippi. This line of watershed undulates between the 52nd and 49th parallels of latitude, from Belle Isle Strait to the sources of the Saskatchewan, in the Rocky Mountains, where it inflects towards the Pacific Ocean, skirting on the north the basin of the Columbia.

Thus bounded on the south, the Arctic lands of America, including the groups of islands lying to the north and north-east, cannot occupy less than 560,000 square leagues. They exceed, therefore, the superficial area of the European lands, estimated at about 490,000 square leagues.

We propose to divide these lands into two zones or regions, the wooded and the desert zones: the former, in America, includes the basins of the Upper Mackenzie, the Churchill, the Nelson, and the Severn.

In the wooded zone the thermometer does not rise above zero until the month of May. Then, under the influence of a more genial temperature, the breath of life passes into the slumbering, inert vegetation. Then the reddish shoots of the willows, the poplars, and the birches hang out their long cottony catkins; a pleasant greenness spreads over copse and thicket; the dandelion, the burdock, and the saxifrages lift their heads in the shelter of the rocks; the sweetbrier fills the air with fragrance, and the gooseberry and the strawberry are put forth by a kindly nature; while the valleys bloom and the hill-sides are glad with the beauty of the thuja, the larch, and the pine.

The boundary between the wooded zone and the barren would be shown by a line drawn from the mouth of the Churchill in Hudson Bay to Mount St. Elias on the Pacific coast, traversing the southern shores of the Bear and the Slave Lakes. To the north, this barren zone touches on eternal snow, and includes the ice-bound coasts of the Parry Archipelago; to the east and the north-east, identity of climate and uniform character of soil bring within it the greatest part of Labrador and all Greenland.

In Asia the isothermal line of 0° descends towards the 55th parallel of latitude, one lower than in America,—though to the north of it some important towns are situated, as Tobolsk, lat. 58° 11′; Irkutsk, lat. 58° 16′; and Yakutsk, lat. 62°.

In Continental Europe, the only Arctic lands properly so called, and distinguished by an Arctic flora, are Russian Lapland and the deeply-indented coast of Northern Russia. Far away to the north, and separated from the continent by a narrow arm of the sea, lie the three almost contiguous islands known as Novaia Zemlaia (lat. 68° 50′ to 76° N.). And still further north, almost equidistant from the Old World and the New, lies the gloomy mountainous archipelago of Spitzbergen (lat. 77° to 81°, and long. 10° to 24°).

We have now only to recapitulate the general characters of the Arctic flora, as they would present themselves to a traveller advancing from the wooded zone into the desert, and thence to the borders of the Polar Sea.

On the southern margin of the wooded region, as in Sweden, Russia, and Siberia, extend immense forests, chiefly of coniferous trees. As we move towards the north these forests dwindle into scattered woods and isolated coppices, composed chiefly of stunted poplars and dwarf birches and willows. The sub-alpine myrtle, and a small creeping honeysuckle with rounded leaves, are met with in favourable situations. Continuing our northerly progress, we wholly leave behind the arborescent species; but the rocks and cliffs are bright with plants belonging to the families of the ranunculaceæ, saxifragaceæ, cruciferæ, and gramineæ. To the dwarf firs and pigmy willows succeed a few scattered shrubs—such as the gooseberry, the strawberry, the raspberry, pseudo-mulberry (*Rubus chamæmorus*)—indigenous to this region, and the Lapland oleander (*Rhododendron laponicum*).

Still advancing northward, we find, at the extreme limits of the mainland, some drabas (*Cruciferæ*), potentillas (*Rosaceæ*), burweeds and rushes (*Cyperaceæ*), and lastly a great abundance of mosses and lichens. The commonest mosses are the *Splechnum*, which resembles small umbels; and, in moist places, the *Sphagnum*, or bog-moss, whose successive accumulations, from a remote epoch, have formed, with the detritus of the *Cyperaceæ*, extensive areas of peat, which at a future day will perhaps be utilized for fuel.

We come now to examine the forms of Animal Life which exist under the conditions of climate and vegetation we have been describing.

Foremost we must place the animal which, in the Arctic World, occupies much the same position as the camel in the Tropical,—the reindeer (*Cervus turandus*).

In size the reindeer resembles the English stag, but his form is less graceful and more compressed. He stands about four feet six inches in height. Long, slender, branching horns embellish his head. The upper part of his body is of a brown colour, the under part is white; but as the animal advances in years his entire coat changes to a grayish-white, and, in not a few cases, is pure white. The nether part of the neck, or dewlap, droops like a pendent beard. The hoofs are large, long, and black; and so are the secondary hoofs on the hind feet. The latter, when the animal is running, make by their collision a curious clattering sound, which may be heard at a considerable distance.

The reindeer anciently invaded Europe and Asia to a comparatively low latitude; and Julius Cæsar includes it among the animals of the great Hercynian forest. Even in our own time large herds traverse the wooded heights of the southern prolongation of the Ouralian range. Between the Volga and the Don they descend to the 46th parallel; and they extend their wanderings as far as the very foot of the Caucasus, on the banks of the Kouma. Still, the proper habitat of the reindeer is that region of ice and snow bounded by the Arctic Circle,—or, more exactly, by the isothermal line of 0° C.

Both the wild and the tame species change their feeding-grounds with the seasons. In winter they come down into the plains and valleys; in summer they retire to the mountains, where the wild herds gain the most elevated terraces, in order to escape the pertinacious attacks of their insect-enemies. It is a fact worthy of note that every species of animal is infested by a parasitical insect. The œstre so terrifies the reindeer that the mere appearance of one in the air will infuriate a troop of a thousand animals. In the moulting season these insects deposit their eggs in the skin of the unfortunate animal, and there the larvæ lodge and multiply *ad infinitum*, incessantly renewing centres of suppuration.

To the natives of North America the reindeer is invaluable. There is hardly a part of the animal not made available for some useful purpose. Clothing made of its skin is, according to Sir J. Richardson, so impervious to cold, that, with the addition of a coverlet made of the same material, any one so protected may bivouac on the snow with safety in the most intense cold of the Arctic night. The venison, when in high condition, has several inches of fat on the haunches, and is said to equal that of the fallow-deer in our English parks; the tongue, and a portion of the tripe, are reckoned most delicious morsels. Pemmican is made by pouring one-third part of

WILD REINDEER.

fat over two-third parts of the pounded meat, and mixing fat and meat thoroughly together. The Eskimos and Greenlanders consider the stomach, or paunch, with its contents, a special delicacy; and Captain Sir James Ross says that the contents form the only vegetable food ever tasted by the natives of Boothia. For the reindeer is a herbivorous animal, and feeds upon the mosses and grasses.

The reindeer is by no means a graceful animal; its joints are large, and powerful in proportion to its size; the divided hoofs are very large, and as the animal is compelled to lift its feet high when going over the snow, its gallop has none of that beautiful elastic spring which characterizes the deer of our own islands, though its pace is "telling," and soon carries it ahead of everything but the long-winded, long-legged wolf.

The stags cast their antlers, and the does drop their young, in May or June, about the time of the first thaw. The males and females are then very seldom found together; the female deer collecting in small herds with their young; the little creatures, which seem all eyes, ears, and legs, taking alarm at any unaccustomed sound or the slightest appearance of danger. The summer vegetation fattens the bucks and does amazingly, and the fawns thrive and develop; all three, says Osborn, having a comparative holiday, and getting into condition to face the trials of the coming winter; while the wolf and the fox, their sworn enemies, are pursuing the infant seals and bears, or attending to their own little domestic duties. But when the autumn frost sets in, and hardens the ground, and the dense snow once more overspreads the dreary northern landscape, the wolves resume their attacks on the unfortunate deer.

For warmth or protection, and following the natural instincts of gregarious animals, they now begin to collect together in large herds of bucks, does, and fawns, numbering as many as sixty and seventy head. The stags seem to undertake the discipline of these large companies, as well as to be responsible for their safety.

Captain Mecham relates that, in October 1852, when crossing that part of Melville Island which intervenes between Liddon Gulf and Winter Harbour, he fell in with as many as three hundred head of deer; and he adds that reindeer were always in sight, in herds varying from ten to sixty in number. One of these herds, containing twenty males, he tried to stalk up to on the 7th of October, but failed in getting a shot at them; for although the does, with the inherent weakness of their sex, showed an excessive curiosity, and made one or two efforts to desert the herd and examine the stranger, the stags would in nowise tolerate such conduct, but chastised them smartly with their antlers, and kept the herd together and in motion by running rapidly round and round, uttering at the same time a strange noise which seemed to alarm the herd, and keep it flying from the suspected danger.

The coat of the reindeer in summer-time is remarkably thin, and adapted admirably in colour to that of the snow-denuded soil; but as winter approaches, it thickens, and gradually resumes its snowy whiteness. Though not, strictly speaking, a *fur*, it forms an admirable non-conducting substance.

As winter, "ruler of the inverted year," extends his sway over the Polar World, and food grows scarce and indifferent, and has to be sought over larger areas, the herds break up into companies of ten or twenty animals; the lichens, the reindeer moss already described (*Cetraria Islandica*), and the sprouts of the creeping willow forming their principal food.

On this branch of our subject Admiral Sherard Osborn makes two suggestive remarks.

Arctic vegetation, he observes, has no time in the autumn to wither or decay—while in full bloom, and before the juices have time to return into the parent root or be otherwise dissipated, the "magic hand of the frost king" strikes them; and thus the wisdom of the Creator has provided for the nourishment of his creatures a fresh and warmth-creating food, lying hid under a mantle of snow, which the instinct of those Arctic animals teaches them to remove and reach the stores so beneficently preserved beneath.

Moreover, most herbivorous animals have a slow system of digestion, even in a domestic state; as, for instance, our cattle and sheep. This appears to be more conspicuously the case in the musk-ox, the reindeer, and the Arctic hare, and is of great utility in lands where vegetation is scanty and widespread, and the weather occasionally so severe as to compel these creatures, for two or three days at a time, to think only of their safety by seeking shelter from the snow-storms in deep ravines or under lofty cliffs. It appears in their case as if Nature extracted from their food a greater quantity of nourishment than she does from that of animals in more southern latitudes; or possibly, the food, by the mere act of remaining in the stomach or intestines, serves to check the cravings of appetite, though no further nutriment should be extracted.

Most of the musk-oxen and deer shot in Captain M'Clintock's expedition, and especially the musk-oxen, had their entrails distended with food apparently quite digested, while the surrounding country in many cases was absolutely barren and lifeless,—inducing the conclusion that these creatures had been a long time collecting their supplies, as also that it had been a long time swallowed, and necessitated the full activity of the vital principle to prevent the food from proving a source of disease. This, indeed, was clearly proved in the case of the musk-oxen, which, if shot, and left twelve hours without being disembowelled, grew tainted throughout with a strong musky odour, rendering the flesh uneatable.

It may also be stated, as an illustration of the facility with which the reindeer can winter in high latitudes, that in Lapland, where they are used as beasts of draught, a daily supply of four pounds of lichen (*Cenomyce rangiferina*) is considered ample for a working animal; and on this dietary a reindeer will be in sufficiently good condition to go without food occasionally for two or three days, and yet, to all appearance, not to be distressed.

Thus, as regards its stores of food, and its provision against the severity of the Arctic winter, the reindeer would seem to be suitably and amply endowed; and its greatest trial is the incessant rapacity of the wolves that follow its track throughout the winter season. As that season advances, the unfortunate animal apparently resigns itself to an evil which it cannot avoid or avert; and the calm composure with which a small troop of these creatures will graze with an *entourage* of half a dozen wolves is not less curious to the observer than philosophical on the part of the reindeer!

"A herd of deer," says an eye-witness, "thus surrounded by the wolves, who were too great cowards to rush in upon their prey, would be startled every now and then by the long-drawn unearthly howl of the hungry brutes; sometimes a frightened deer, horror-stricken at the abominable chant, dashes madly away from the herd,—away all, or a portion, of the wolfish fraternity go after it. In many cases the scene may be briefly summed up with the old three-volume *denouement* of—a rush, a shriek, a cranching of bones, and snarling of beasts of prey, and all is over! for the wonderful powers of swallow and horrid voracity of an Arctic wolf must be seen to be understood; no writer would peril his reputation for veracity by repeating what has been seen on that head. But sometimes the frightened deer gains the open country, and goes wonderful distances dogged by the persevering wolf, who assuredly has it, unless another herd is met which admits the hunted deer into its ranks.

"Occasionally, whilst a herd of deer are grazing, one of them may happen to hit upon a spot where the food is plentiful; it naturally lingers there, while the herd is moving slowly on against the wind. The wolves immediately mark the straggler, and stealthily crawl on, their object being to cut him off from the herd; that effected, there is a howl and a rush, which if the deer does not evade by extraordinary exertions, his fate is instantly sealed."

These scenes are enacted throughout the long Arctic winter. When sight is rendered useless, scent comes to the aid of the rapacious destroyer; and we can well believe that many an explorer, in the December darkness of the frozen wastes, has often wished his olfactory nerves were as sensitively organized as those of the wolf. For although he can then hear the reindeer, it is impossible to see them, except when they hurry across the dark but snowy landscape; and many a bad shot has been made by a hungry seaman at a large pair of melancholy eyes which peered out of the enveloping mist, because he could not tell, for the life of him, whether the animal was distant two or twenty yards.

In the dreadful winter of 1852–53, the deer approached close to the exploring-ship *Investigator*, having quitted the land and traversed the belt of ice. It is difficult to say whether this was done with a view of seeking the warmth which instinct, if not scent, told them radiated from the vessels,—the vessels, compared with the temperature everywhere prevailing (namely, 9° 5' below freezing-point), being complete volcanoes of heat; or whether it was for security against their wolfish enemies. Probably, it was for the first-named reason; inasmuch as it is recorded that the foxes of Leopold Harbour, in 1848, soon became aware of the warmer atmosphere produced by the presence of Sir James Ross's squadron, and sagaciously burrowed and bred in the embankments thrown up around the ships.

But, at length, winter and its sorrows pass away, and early in the new year a happier life dawns on the much-tried reindeer. In February and March the seals begin to breed, and as the attention of the wolves and other beasts of prey is then drawn to the helpless young, which are truly "delicious morsels," the holidays of the reindeer may be said to commence. We may remind the reader also that the Arctic hare and the lemming winter in the icy north, and yield occasional meals to wolf and fox.

The spring returns, and as the sun rises above the horizon, the great herds gradually break up and scatter abroad; and the deer may then be seen in wandering groups of three or four, until once more the autumn-twilight deepens, and they reassemble in numerous companies.

As the reindeer is the camel of the Polar World, so the Arctic wolf may be said to occupy the place of the tiger; so daring is its courage, and so fierce its lust of blood. Assembling in large packs, they are not afraid to haunt the immediate neighbourhood of man. In Captain M'Clintock's expedition, they gathered round the *Investigator* at such close quarters, that it was unsafe for the crew to leave the ship, unless in companies, and well-armed; and with their melancholy howls they made night hideous. Five of them attempted to pounce on an Eskimo dog which had long been the pet of the *Investigator*. One of these brutes is described as a "perfect giant," standing nearly four feet high at the shoulder, and having a footmark as big as a reindeer's.

Our English seamen planned many a clever scheme to entrap these wary creatures, but all failed, while some of the encounters with them were unpleasantly close, and the risk very considerable. One day, the boatswain, while out shooting, broke by a shot two of the legs of a fine buck reindeer. Evening coming on, and he knowing the animal could not drag itself far, returned to the ship. Next morning, he started at an early hour to secure his prize. What was his disgust, when he arrived at the place, to find his booty in the possession of five large wolves and several foxes! Determined to have, at all events, a share, the boatswain advanced, shouting with all his might, and hurling at the thieves every opprobrious phrase he could invent, yet afraid to fire his single-barrelled gun at any one of them, for fear the rest should serve him as they were serving the buck; more particularly as they appeared inclined to show fight, and made no sign of retreat until he was within four yards. Even then only four had the grace to move away, sitting down a pistol-shot off, and howling most lamentably.

The boatswain picked up a leg of the deer, which had been dismembered, and then grasped one end of the half-devoured carcass, while a large she-wolf tugged against him at the other.

It must be owned that this position was a disagreeable one, and had the howling of the four wolves brought others of their kind to the rescue, the consequences of this affray between hungry wolves and a no less hungry sailor might have proved serious. Fortunately, the interpreter, who chanced likewise to be out shooting on a neighbouring hill, had his attention attracted by the noise of the brutes, and made his appearance on the scene. He afterwards described it as the strangest he had ever witnessed. So close were the boatswain and the carnivora in their struggle for the meat, that he fancied the latter had actually attacked the former. On the arrival of this reinforcement the wolves decamped, leaving the gallant boatswain with only twenty pounds weight of meat, instead of the one hundred and twenty his prize must have originally weighed.

The identities between the Arctic dog and the Arctic wolf are so important that Dr. Kane agrees with Mr. Broderip in assigning to these animals a family origin. The oblique position of the wolf's eye is not uncommon among the Eskimo dogs. Dr. Kane had a slut, one of the tamest and most affectionate of his team, who had the long legs, the compact body, the drooping tail, and the wild scared expression of the eye, which some naturalists have supposed to distinguish the wolf alone. When domesticated early—and it is easy to domesticate him—the wolf follows and loves you like a dog. "That they are fond of wandering proves nothing; many of our pack will stray for weeks," says Kane, "into the wilderness of ice; yet they cannot be persuaded, when they come back, to inhabit the kennel we have built for them only a few hundred yards off. They crouch around for the companionship of men." Both animals howl in unison alike; and, in most parts, their footprint is the same.

The musk-ox (*Ovibos moschatus*) is one of the largest of the Polar ruminants. As its zoological name indicates, it is an intermediary between the ox and the sheep. Smaller than the former, larger than the latter, it reminds us of both in its shape and general appearance. It has an obtuse nose; horns broad at the base, covering the forehead and crown of the head, and curving downwards between the eye and ear until about the level of the mouth, where they turn upwards; the tail is short, and almost hidden by the thickness of the shaggy hair, which is generally of a dark brown, and of two kinds, as with all the animals of the Polar Regions; a long hair, which on some parts of the body is thick and curled, and, underneath, a fine kind of soft, ash-coloured wool; the legs are short and thick, and furnished with narrow hoofs, like those of the moose. The female is smaller than the male, and her horns are smaller. Her general colour is black, except that the legs are whitish, and along the back runs an elevated ridge or mane of dusky hair.

The musk-ox, as his name implies, throws out a strong odour of musk,—with which, indeed, his very flesh is impregnated, so that the scent is communicated to the knife used in cutting up the animal. Not the less is he regarded as a valuable booty by the Indians and the Eskimos, who hunt him eagerly. He wanders in small troops over the rocky prairies which extend to the north of the great lakes of North America. He is a fierce-tempered animal, and in defence of his female will fight desperately.

His general habits resemble strongly those of the reindeer; but his range appears to be principally limited to Melville Island, Banks Land, and the large islands to the south-east of the latter.

One of our Arctic explorers describes the musk-oxen as all very wild in April, and as generally seen in large herds from ten to seventy in number. In June they were stupidly tame, and seemed to be oppressed by their heavy coats of wool, which were hanging loosely down their shoulders and hind-quarters in large quantities; the herds much smaller, and generally composed of cows and calves.

The heavy coat of wool with which the musk-oxen are provided, is a perfect protection against any temperature. It consists of a long fine black hair, and in some cases white (for it is not ascertained that these oxen change their colour during the winter), with a beautiful fine wool or fur underneath, softer and richer than the finest alpaca wool, as well as much longer in the staple. This mantle apparently touches the ground; and the little creature looks, it is said, like a bale of

THE MUSK-OX.

black wool, mounted on four short nervous goat-like legs, with two very bright eyes, and a pair of sharp "wicked-shaped" horns peering out of one end of it.

They seem to be of very uncertain temper, sometimes standing stupidly glaring at their assailants, whetting their horns against their fore legs; at other times, they will rush furiously against their hunters.

Captain Mecham discovered very great numbers of musk-oxen near the head of Hardy Bay, Melville Island. On one plain he observed as many as seventy grazing within a circuit of two miles; on his approach, they divided into herds of about fifteen each, headed by two or three enormous bulls. Their manœuvres, he says, were so quick and regular that they might be more fittingly compared to squadrons of cavalry than anything else he could think of. One herd moved forward at a gallop, several times within rifle-shot, and formed in perfect line with bulls in the van, presenting a formidable array of horns. The last time they advanced at a gallop until within about sixty yards, when they formed in line, the bulls snorting wildly, and tearing up the snow. But as soon as Captain Mecham fired they wheeled round promptly, rejoined the main body, and made off out of sight, only waiting occasionally for the wounded animal.

The following graphic account of an encounter with a musk-ox is given by Captain M'Clintock:—

"We saw and shot two very large bulls—a well-timed supply, as the last of the venison was used up; we found them to be in better condition than any we had ever seen. I shall never forget the death-struggle of one of these noble bulls; a Spanish bull-fight gives no idea of it, and even the slaughter of the bear is tame in comparison. This animal was shot through the lungs, and blood gushed from his nostrils upon the snow. As it stood fiercely watching us, prepared yet unable to charge, its small but fixed glaring eyes were almost concealed by masses of shaggy hair, and its whole frame was fearfully convulsed with agony; the tremulous motion was communicated to its enormous covering of tangled wool and hair; even the coarse thick mane seemed to rise indignant, and slowly waved from side to side. It seemed as if the very fury of its passion was pent up within it for one final and revengeful charge. There was no roaring; the majestic beast was dumb; but the wild gleam of savage fire which shot from his eyes, and his menacing attitude, were far more terrible than the most hideous bellow. We watched in silence, for time was doing our work, nor did we venture to lower our guns until, his strength becoming exhausted, he reeled and fell.

"I have never witnessed such an intensity of rage, nor imagined for one moment that such an apparently stupid brute, under any circumstances of pain and passion, could have presented such a truly appalling spectacle. It is almost impossible to conceive a more terrific sight than that which was presented to us in the dying moments of this matchless denizen of the northern wilds."

It seems doubtful whether the wolf, which is naturally a most cowardly creature, can act on the offensive against the musk-ox; and most Arctic navigators seem of opinion that it attacks only lame or sickly cattle.

The activity of these oxen, and their goat-like power of climbing, is very remarkable, and much at variance with their clumsy appearance. They have been seen making their way, when frightened, up the face of a cliff which defied all human efforts, and going down the precipitous sides of ravines by alternately sliding upon their hams, or pitching and arresting their downward course, as Sherard Osborn remarks, by the use of the magnificent shield of horn which spreads across their foreheads, in a manner to excite the liveliest astonishment of the spectator.

The Arctic Fox (*Canis lagopus*) cannot compare with either of the preceding animals in importance or interest, yet it figures very largely in the journals of our Arctic explorers. It is smaller than the common European fox; has a sharp nose, and short rounded ears, almost concealed in its fur; the legs are short, and the toes covered both above and below with a thick soft fur; the tail is shorter than that of the common fox, but more bushy. Its range is very extensive, for it is found in the lands bordering on the Polar Sea in both continents. As winter approaches, its coat of hair grows thick and ragged; until at length it becomes as white as snow—the change of colour taking place last on the ridge of the back and the tip of the tail. Its food consists of various small quadrupeds,—such as the Arctic hare and the lemming,—on all kinds of water-fowl and their eggs, on the carcasses of fish, shell-fish, and the refuse of the young seals killed and devoured by the Polar bear. In the track of the latter it seems to hunt systematically. It swims with dexterity, and will cross from island to island in search of prey. Its fur is light and warm, though not very durable, and for the sake of this fur it is pursued both in Arctic Asia, Greenland, and Hudson Bay. It is a wary animal, however, and not easily caught.

ARCTIC FOXES.

Dr. Hayes affords us an illustration of this statement.

As he and a follower, named Bonsall, on one occasion were exploring in Northumberland Island, they discovered a fox scampering away over the plain. Bonsall gave chase, but could not arrive within shooting distance. Another was then heard barking overhead at them. Dr. Hayes seized his gun, and climbing over some huge boulders which filled the bottom of the gorge, endeavoured, by crawling behind a rock, to overtake or approach the animal; but it seemed to be aware of his intentions, and scampering away, led him a wild chase across the plain. The astute Reynard first made off, so that his assailant "could not cover him upon the cliff;" and when out of danger, perched upon a stone, and barked at him in the most tantalizing manner. The doctor approached within long range. Immediately, as he was about to bring his gun to his shoulder, it dropped behind the stone and fled to another, where it set up the same rapid chatter,—a shrill "Huk! huk! huk!" sounding like a mixture of anger and defiance. Again Dr. Hayes tried to approach it, but with no better success; round and round it ran, until at length, weary of following it, Dr. Hayes fired. Some of the shot probably touched it, for it screamed loudly; but it fled with remarkable rapidity, and finally baffled its pursuer.

As the flesh of the fox is by no means to be despised, and, indeed, ranks as a dainty in the bill of fare of an Arctic navigator, a hot pursuit of it is often maintained, and traps are constructed to ensure its capture. These are usually built on much the same principle as a rabbit-trap. Selecting a smooth level rock, the trappers arrange some flat stones of about six inches thick, so as to enclose on three sides an area of six inches by two feet and a half. Over this enclosure other flat stones are laid; and between the two used to close up one of the ends a peg is inserted, so as to project about an inch within the trap.

To this peg, by means of a loop, is loosely hung a small piece of meat; and to the same peg, outside, is attached another loop made at the end of a cord, the cord being carried up through the rear of the trap, and over the top to the front, where it is fastened round a thin flat flag of slate, which moves freely up and down, being guided and held by a couple of large blocks placed one on either side of the entrance.

The way in which this machinery works is very simple:—

The fox enters under the slide or trap-door, advances to the rear, seizes the bait, and attempts to back out. The bait, of course, is pulled from the peg, and with it the loop supporting the door comes off. As soon as its support is removed down comes the door, and Master Reynard is entrapped. Everything now depends on the manner in which the cracks have been closed up; for if the animal can thrust its little nose between a couple of stones, it will assuredly effect its escape. Nor is it less important that the enclosure should not be sufficiently large to enable the fox to turn round; for in that case it generally contrives to loosen the door, and depart in infinite glee.

A FOX TRAP.

The Arctic fox is described by Dr. Hayes as the prettiest and most provoking of living creatures. One which he unsuccessfully chased for fully three hours was about the size of a domestic cat, round and plump, white as the snow, with a long pointed nose, and a trailing bushy tail, which seemed to be its particular pride. It was quite evident that it enjoyed the perplexities of its hunters, as it leaped from rock to rock, or circled round and about them, and showed the utmost indifference to the miseries of their famished condition. It rolled and tossed about among the loose drift, now springing into the air, now bounding away, now stopping short, and now cocking its head to one side and elevating one foot, as if listening, seeming all the time to be intent on exhibiting its "points" to its enemies, for whom it did not care the value of the minutest part of its very pretty tail. Weary and exhausted, Dr. Hayes abandoned the pursuit, and returned to his camp, followed by the fox, though always at a safe distance; and when they last caught sight of it, as they looked back from the rocks above the hut, it was mounted on an elevation, uttering its shrill sharp cry, in apparent mockery of their defeat.

Of the supposed relations between the bear and the fox, Dr. Kane remarks that he once thought his observations had confirmed them. It is certain that they are frequently found together; the bear striding on ahead with his prey, the fox behind gathering in the crumbs as they fall; and Dr. Kane often saw the parasite licking at the traces of a wounded seal which his champion had borne off over the snow. The story is that the two hunt in couples. This may well be doubted, though it is clear that the inferior animal rejoices in his association with the superior, at least for the profits, if not the sympathy it brings to him. "I once wounded a bear," says Dr. Kane, "when I was out with Morton, and followed him for twelve miles over the ice. A miserable little fox travelled close behind his patron, and licked up the blood wherever he lay down. The bear at last made the water; and as we returned from our fruitless chase, we saw the fox running at full speed along the edge of the thin ice as if to rejoin him."

A welcome addition to the meagre fare of the Arctic navigator is furnished by the Arctic Hare (*Lepus glacialis*), which, like the reindeer, collects in herds or troops as winter approaches. As many as two hundred have been seen at a time; and at one of their favourite haunts—Cape Dundas, Melville Island—might be seen a complete highway, three yards broad, which the tread of their numbers had beaten through the snow. In winter they seek their food and burrow for protection under the snow-crust. Captain M'Clintock states that they are ubiquitous in the Polar Regions, but that, of course, they are most numerous where the pasture is most abundant, as on Banks Land and Melville Island. The sportsmen of the two discovery ships, *Resolute* and *Intrepid*, shot one hundred and sixty-one hares in a twelvemonth on Melville Island; their average weight, when fit for the table, was seven pounds, and from ten to twelve pounds including skin and offal.

In the warm brief summer the hare takes refuge from the pursuit of beasts of prey under large boulders, or in the steep face of rocky ravines. It is then found in groups of from twelve to twenty. So delicate is their skin, that though the winter fur is of exceeding beauty and brilliant whiteness, it cannot be applied to any purpose of utility. They do not hibernate; and our explorers generally found them amongst the heavy hummocks of the floe-ice, as if they fled to that rugged ground from the wolves or foxes.

In the range of the Altaï, and extending even into Kamtschatka, we meet with the Alpine Hare (*Lagomys Alpinus*); a small rodent, scarcely exceeding a guinea-pig in size, and measuring in length nine inches only: it has a long head, with short, broad, and rounded ears. Its favourite places of sojourn are among the rocks and cataracts of wild wooded regions, where it forms burrows beneath the rocks, or inhabits their fissures. When the sky is bright, and the sunshine genial, they seldom leave their holes in the day-time; but in dull weather they may be seen bounding among the rocks, and making the echoes resound with their low whistle or bird-like chirp. In the autumn they make ready against winter need by collecting a large assortment of the most nutritious herbs and grasses, which, after drying in the sun, they arrange in heaps of various sizes, according to the number of animals engaged in the task; and as these heaps are often several feet in height and breadth, they may be easily distinguished even through the deep snow, and frequently prove of great service to the Siberian sable-hunters, whose horses would perish but for the supplies thus strangely afforded. Hence, wherever a Siberian or a Tartar tribe is found, the Alpine hare possesses a distinctive name,—and, notwithstanding its diminutiveness, is highly valued.

Another rodent which deserves to be remembered in these pages, is the Arctic or Hudson Bay Lemming (*Myodus lemmus*), which is found in Labrador, and on all the American mainland washed by the cold waters of the Arctic Ocean. It has been described as "a perfect diamond edition of the guinea-pig." In habits it resembles the hare very closely, except that it is more gregarious, and is generally found in large families. In summer it is of an ashy colour, with a tawny tinge on the back, a dusky streak along its middle, and a pale stripe on either side. It has the repute of being exceedingly inoffensive; and is tamed so easily that, when caught, it becomes reconciled to its captivity in a day or two, and will soon show itself sensible of its master's caresses. In winter it is perfectly white,—white as snow, from which it can be distinguished only by the keen scent of the fox or the Eskimo dog.

About the end of May, or early in June, it leaves the land and seeks the floating ice; for what purpose does not seem as yet to be accurately ascertained. Is it due to an instinct of migration, such as the Norwegian lemming so powerfully exhibits? It may be that the thaws force them from the land, or that, as the seamen say, "Them blessed little lemmings must be arter salt!" They have often been found steering off shore from the north coast of Melville Island, leaving comparative plenty in their rear; and, so far as could be made out, on a clear day, from land of considerable height, there was nothing in the shape of *terra firma* in the direction they were taking. When thus exposed upon the open floe, owls, gulls, and foxes pick them up for food. Can it be that Providence occasions, or has ordained this exodus for the purpose of feeding these creatures, and of thinning down the numbers of an animal which would otherwise multiply exceedingly, and devour all the vegetation of a naturally barren region?

From an Arctic journal it would appear that the lemmings are preyed upon by the Polar bear We transcribe a graphic passage in further illustration of the habits of that remarkable carnivore:—

"Seeing some drift-wood lying about," says a gallant navigator, "which it was important should be examined, I halted and encamped, dispersing the men along the beach to bring all in they could find. Walking landward to obtain a view from a hill, I was startled to see a she-bear and two cubs some distance inland. Watching them carefully, I was not a little interested to see the mother applying her gigantic muscular power to turning over the large blocks of sandstone which strewed the plain, and under which the unlucky lemmings at this season take shelter. Directly the she-bear lifted the stones, which she did by sitting upon her hams and pulling them towards her with her fore paws, the cubs rushed in and seized their prey, tossing them up in the air in their wantonness. After repeating this operation until the young fry must have made a very good meal, I was glad to witness the bear's mode of suckling her young—a sight, I should think, rarely seen. Seated on her haunches, with the backbone arched, so as to bring the breasts (which were situated between the shoulders) as low as possible, the youngsters sucked away in a standing attitude. Anxious to secure this family-party, we proceeded to burn all sorts of strong-smelling articles; and at last she brought her babes down, though very warily, and when more than one hundred yards off turned away, evidently suspicious."

In the sub-Arctic regions are found some of those animals which furnish commerce with the costliest furs. They all belong, however, to the family *Mustelidæ*, represented in temperate climes by the common weasel (*Musteles*).

The marten of North America is, in fact, the cousin-german of the weasel, and not less ferocious in its habits. In the forests of fir and birch which it loves to frequent, it preys upon the small rodents, the birds, and, if its appetite is very keen, upon the reptiles. It scales trees as nimbly as the cat; and its flexible body enables it to insinuate itself into the smallest openings, where a cat could not pass, and into the burrows and hollows of the trees or rocks in which its victims seek shelter. It is, however, a pretty animal, with vivacious ways, an astute physiognomy, and a rich coat of fur.

In the wooded zone which borders on the desert region of the Polar World are found both the Pine Marten (*Mustela martes*) and Pennant's Marten (*Mustela Canadensis*). The fur of the former is of a very superior quality, and its skin forms a great article of commerce. It burrows in the ground, and feeds upon mice, rabbits, and partridges. The Canadian marten is larger than the preceding; longer and stronger. It lives in the woods, preferring damp places to dry; and climbs with a remarkable amount of ease and dexterity.

The Sable (*Mustela zibellina*) is much more highly esteemed for its fur than any other of the weasel tribe. It has long whiskers, rounded ears, large feet (the soles of which are covered with fur), white claws, and a long bushy tail. The general colour of the fur is brown, more or less brilliant, but the lower parts of the throat and neck are grayish.

THE ERMINE AND SABLE MARTEN.

A vivacious and nimble animal is the sable, which dwells in the remotest recesses of the forests, beneath the roots of trees, and in holes of the earth, and penetrates to the very borders of the realm of perpetual snow. Prodigious numbers are killed in Siberia, during the months of November, December, and January. The hunters assemble in large companies, and make their way down the great rivers in boats, carrying sufficient supplies of provisions for a three months' absence. On reaching the appointed place of rendezvous, the different companies, each under the direction of a leader, fix upon their respective quarters, erect huts of trees, and build up the snow around them. In the neighbourhood of these they lay their snares; and then, advancing another mile or so, they set a further quantity; and thus they proceed, until they have covered a considerable area of ground; building huts in each locality, and returning in due order to each set of snares to collect their prey. These snares are of the simplest construction; nothing more than small pits or cavities, loosely covered with rough planks or branches of trees, and baited with fish or flesh. When the game grows scarce, the trappers follow the sable to their retreats by tracking their footprints over the fresh-fallen snow; place nets at their entrances; and quietly wait, if it be for two or three days, until the animals make their appearance.

The fur of the sable is distinguished from all other furs by this singular property: the hair has no particular inclination, but may be laid down indifferently in any direction whatever.

The genus Polecat (*Mustela putarius*) comprehends the smallest of all known carnivores,—namely, the weasel, the ferret, and the ermine. The temperate countries of Europe possess a variety of the latter species; but the ermines of the remote North yield the fullest and softest fur. These animals, like many others in high latitudes, change the colour of their coat according to the season. They have been adopted by poets, on account of the spotless whiteness of their fur, as emblematic of purity; but, in truth, they merit that honour only in the winter: in the summer their colour is a clear maroon. The tail, at all times, is of a beautiful brilliant black.

Another carnivorous quadruped which haunts the northern forests is the Glutton (*Gulo Arcticus*), or Wolverine; it owes its former and more popular name to its extreme voracity. But it is at least as remarkable for its strength and fierceness, inasmuch as it does not fear to dispute their prey with the wolf and bear; and for its cunning, since it baffles again and again the most carefully devised stratagems of the hunter. It is a slow and somewhat unwieldy

THE GLUTTON, OR WOLVERINE.

animal; but it is determined and persevering, and will proceed at a steady pace for miles in search of prey, stealing unawares upon hares, marmots, and birds; and surprising even the larger quadrupeds, such as the elk and the reindeer, when asleep.

The stories told of this remarkable animal's shrewdness, which far exceeds that commonly attributed to the fox, would seem incredible, were they not confirmed by good authority. It is in allusion to its extraordinary cunning that the Indians call it *Kekwaharkess*, or the "Evil One." With an energy that never flags it hunts day and night for the trail of men, which, when found, it follows up unerringly. On coming to a lake, where the track is generally drifted over, it continues its steady gallop round the shores, to discover the point at which the track re-enters the woods, when it again pursues it until it arrives at one of the wooden traps set for the marten or the mink, the ermine or the musk-rat. Cautiously avoiding the door, it effects violent entrance at the back, and seizes the bait with impunity; or, if the trap contains an animal, drags it out, and, with wanton malevolence, mauls it, and hides it at some distance in the underwood or at the top of some lofty pine. If hard pressed by hunger, it devours the victim. And in this manner it demolishes the whole series of traps; so that when once a wolverine has established itself on a trapping-walk, the hunter's only chance of success is to change his ground, and build a fresh lot of traps, in the hope of securing a few furs before the new path is discovered by his industrious enemy.

Some interesting particulars of the habits and ways of the glutton are recorded by Lord Milton and Dr. Cheadle in their lively narrative of an expedition from the Atlantic to the Pacific ("The North-West Passage by Land"). They tell us that it is never caught in the ordinary pit-fall. Occasionally one is poisoned, or caught in a steel trap; but so great is the creature's strength, that many traps strong enough to hold securely a large wolf will not retain the wolverine. When caught in this way, it does not, like the fox and the mink, proceed to amputate the limb, but, assisting to carry the trap with its mouth, hastens to reach a lake or river, where its progress will be unimpeded by trees or fallen wood. After travelling to a sufficient distance to be safe from pursuit for a time, it sets to work to extricate its imprisoned limb, and very frequently succeeds in the attempt.

Occasionally the glutton is killed by a gun placed so as to bear on a bait, to which is attached a string communicating with the trigger. But a trapper assured Lord Milton and his companion, that very often the animal had proved too cunning for him, first approaching the gun and gnawing in two the cord communicating with the trigger, and then securely devouring the bait.

In one instance, when all the trapper's devices to beguile his enemy had been seen through, and clearly foiled, he adopted the plan of placing the gun in a tree, with the muzzle pointing vertically downwards upon the bait. This was suspended from a branch, at such a height that the animal could not secure it without jumping; and, moreover, it was completely screened by the boughs. Now, the wolverine's curiosity almost equals its voracity. It shows a disposition to investigate everything; an old moccasin flung aside in the bushes, or a knife lost in the snow, must be ferreted out and examined, and any object suspended almost out of reach generally proves irresistible as a temptation. In this instance, however, the caution of the glutton exceeded its curiosity and restrained its hunger; it climbed the tree, cut the fastenings of the gun, which then tumbled to the ground, and, descending, it secured the bait with impunity.

Lord Milton's party were personal sufferers by, and witnesses to, the animal's cunning. One day, when setting out to visit their traps, they observed the footprints of a very large wolverine which had followed their trail, and La Ronde, their trapper, at once exclaimed, "C'est fini, monsieur; il a cassé toutes nôtres étrappes, vous allez voir;" and so it proved. As they came to each in succession, they found it broken open at the back, and the bait taken; and, where an animal had been caught, it was carried off. Throughout the whole line every one had been demolished; and the tails were discovered of no fewer than ten martens, the bodies of which had apparently been devoured by the hungry and astute wolverine.

With one more illustration we must be content, and turn to another branch of our subject, though we do not suppose that our readers will weary of the relation of facts which throw so vivid a light on the intelligence, as distinct from, and superior to, the instinct of animals. And, certainly, the manner in which the glutton foils the ingenious stratagems of the trappers must be ascribed to intelligence rather than to instinct. In the following anecdote we think it is plainly shown that the latter could not have sufficed to guard the animal against the machinations of its persevering foes.

Dr. Cheadle, accompanied by an Indian boy, named Misquapasnayoo, started off for the woods, bent upon proving his superior acuteness to the wolverine. They found that the latter had renewed his visits along the line of traps, and broken all which had been reconstructed, devouring the animals found in them. Dr. Cheadle thereupon adopted a device which could not fail, he thought, to catch his enemy in his own toils. All the broken traps were repaired and set again, and poisoned baits substituted for the ordinary ones in the traps; not in every instance, but here and there along the line.

The forest was here of great extent, and seemed to stretch away to the frozen North without let or hindrance, the mass of timber being broken only by numerous lakes and swamps, or clearances which had been caused by conflagrations. The traveller always seeks the lakes; not only because they enable him to travel more rapidly, and penetrate further into the less hunted regions, but also because the edges of the lakes, and the portages between them, are favourite haunts of the fox, the fisher, and the mink. On one of these lakes a curious circumstance was noted. The lake was about half a mile in length, and of nearly equal breadth, but of no great depth. The water had seemingly frozen to the bottom, except at one end, where a spring bubbled up, and a hole of about a yard in diameter existed in the ice-crust, which was there only a few inches thick. In this hole the water was crowded with myriads of small fish, most of them not much larger than a man's finger, and so closely packed that they could not move freely. On thrusting in an arm, it seemed like plunging it into "a mass of thick stir-about." All around the snow had been trodden down hard and level by the feet of the numerous animals attracted to this Lenten banquet; and tracks converged to it from every side. The footprints could be recognized of the cross or silver fox, delicately impressed in the snow as he trotted daintily along with light and airy tread; the rough marks of the clumsier fisher; the clear and sharply-defined track of the nimble mink; and the great cross trail of the ubiquitous glutton. On the trees around scores of crows were sleepily digesting their abundant meals.

When Dr. Cheadle and his companion turned homewards, they found that their enemy had been in active pursuit. Along the ground they had traversed on the previous day, every trap was already demolished, and all the baits were abstracted. Dr. Cheadle at first imagined that he had at last outwitted and destroyed his enemy; but the Indian's keener eyes discovered each of the baits which had been poisoned, lying close at hand, bitten in two and rejected, while all the others had disappeared. The baits, nevertheless, had been very carefully prepared; the strychnine being inserted into the centre of the meat by a small hole, and when frozen it was impossible to distinguish them from the harmless ones by any peculiarity of appearance. It seemed as if the animal suspected poison, and bit in two and tasted every morsel before swallowing it. The baits had purposely been made very small, so that in the ordinary course they would have been swallowed whole. That the same wolverine had followed up their path from the first, they knew perfectly well, because it was one of unusually large size, as shown by its tracks, which were readily distinguished from those of smaller animals.

The distribution of Birds in the Polar Regions, is a subject on which it seems desirable to offer a few remarks, so that our readers may be able to form an accurate conception of the character and variety of the animal life peculiar to them.

Of the birds of Greenland and Iceland, it may be affirmed that fully three-fourths of the species, and a still larger proportion of individuals, are more or less aquatic, and many of the remainder are only summer visitors. The largest bird that ventures far north is the *Aquila albicilla*, or fishing-eagle, which builds its eyrie on the loftiest crags of the ocean-cliffs, and feeds on salmon and trout. The *Falco Islandicus*, or gyrfalcon, though a native of Iceland, is now very rarely met with. The snowy owl inhabits the glaciers which fill the deep inland valleys of Greenland, and its range extends as far southwards as the Orkneys. Particular kinds of grouse are confined to the high latitudes; and more particularly the ptarmigan, or white grouse,

PTARMIGAN.

which supplies a welcome addition to the scanty bill of fare of the Arctic navigators. It is found, even in the depths of winter, on Melville Island; burrowing under the snow, perhaps, for warmth, protection, and food. But it appears to be most numerous in April, when it is found in pairs; in September it collects in coveys, sometimes of as many as fifteen or twenty birds, preparatory to their southern migration.

Of the *Corvidæ*, the only species which ventures beyond the Arctic Circle is the Royston crow, and that only in summer.

The raven, however, is found in all the wide Polar realm, and is larger, stronger, and more voracious in the Arctic Islands than elsewhere. It drives the eider-ducks from their nests in order to prey on their young or feast on their eggs, and it unites in flocks to expel intruding birds from their abode.

The *Grallatores* are more numerous than land-birds in the Arctic World. The snipe and the golden plover are only visitors; but the oyster-catcher is a denizen of Iceland, where, building its nest on the reedy banks of the streams, it wages war with the crow tribe. The heron, curlew, plover, and most of the other waders, emigrate; sand-pipers and the water-ousel remain "all the year round."

The *Cygnus musicus*, or whistling swan, is specially famous for its migrations. It measures five feet from the tip of the bill to the end of the tail, and eight feet across its noble extended wings; its plumage is white as snow, with a slight tinge of orange or yellow on the head. Some of these swans winter in Iceland; and in the long Arctic night their song, as they pass in flocks, falls on the ear of the listener like the notes of a violin.

The distribution of animals is, of course, regulated by laws analogous to those which regulate the distribution of plants, insects, birds, and fishes. Each continent, and even different portions of the same continent, are the centres of zoological families, which have always existed there, and nowhere else; each group being almost always specifically different from all others. As the Arctic World includes a district common to Europe, Asia, and America, with uniform climatic conditions, the animals inhabiting the high latitudes of these continents are frequently very similar and sometimes identical; and, in fact, no genus of quadrupeds exists in the Arctic regions that is not found in all three continents, though there are only twenty-seven species common to all, and these mostly fur-bearing animals. The carnivores, as we have seen, are very few in number, and of these the most important is the Polar bear. Of the herbivores the reindeer is the most valuable; its southern limit in Europe is the Baltic Sea, in America the latitude of Quebec.

There are fully eight varieties of American dogs, several of which are natives of the far North. The *lagopus*, or isatis, a native of Spitzbergen and Greenland, extends over all the Arctic regions of America and Asia, and is found in some of the Kurile Islands. Dogs are employed to draw sledges in Newfoundland and Canada; and the Eskimo dogs, used for this purpose by the Arctic explorers, are famous for their strength, their docility, and power of endurance. They were mute, until they learned to bark from European dogs on board the discovery ships.

CHAPTER VI.

ICELAND AND THE ICELANDERS.

JUST within the Arctic region, but nearly on the limits of what geographers call the Atlantic Ocean, lies an island which, since its colonization in the ninth century, has not ceased to excite the interest of the explorer and the man of science.

Iceland—which measures about 300 miles at its greatest length, from east to west, and about 200 miles at its greatest breadth, from north to south—is situated in lat. 63° 23′–66° 33′ N., and long. 13° 22′–24° 35′ W.; at a distance of 600 miles from the nearest point of Norway, 250 from the Faröe Isles, 250 from Greenland, and above 500 miles from the northern extremity of Scotland. As early as the eighth Christian century it was discovered by some European emigrants; though, indeed, the Landnana Book, one of the earliest of the island-records, asserts that they found the memorials of a yet earlier settlement in various Christian relics, such as wooden crosses, which appeared to be of Irish origin. At all events, the first really successful attempt at colonization was made by Ingolf, a Norwegian, who planted himself and his followers at Reikiavik in 874. In the following century a somewhat extensive immigration took place of Norwegians who resented the changes of polity introduced by Harold Haarfager, and all the habitable points on the coast were occupied by about 950 A.D. Fifty years afterwards, though not without much opposition, Christianity was legally established, and the bishoprics of Holar and Skalholt were founded. The government assumed the character of an aristocratic republic, with a popular assembly, called the Althing, meeting every summer in the valley of Thingvalla. Commerce was encouraged, and the Icelanders early distinguished themselves by the boldness of their maritime enterprise, and the extent of their ocean fisheries.

About the year 932 they discovered Greenland, and about 986 a portion of the North American coast, which they called "Vineland." They did not confine their voyages to the north, but sent their ships even as far south as the Mediterranean. From 1150 to 1250 is rightly considered the most flourishing period of Icelandic literature and commerce. After the conquest of the island by Haco VI. of Norway, much of the old spirit seemed to die out. When Norway was united to Denmark in 1380, Iceland was included in the bond, and it is still regarded as a dependency of the latter kingdom. In 1540 it embraced the principles of Lutheran Protestantism. Its population at one time numbered 100,000, but it gradually diminished until, in 1840, it was reduced to 57,094; but a slow increase has taken place of late years, and it now amounts to about 70,000. The language spoken is the old Norse.

Iceland is a fifth part larger than Ireland, and its superficial area is estimated at 39,207 square miles. Not more than 4000 miles, however, are habitable, all the rest being ice and lava; for the island seems to be little more than a mass of trachyte, snow-shrouded and frost-bound, resting on a sea of fire. It consists of two vast parallel table-lands, the foundations of ranges of lofty mountains, most of which are active volcanoes; and these table-lands strike across the centre of the island, from north-east to south-west, at a distance from one another of ninety to one hundred miles. Their mountainous summits are not pyramidal, as is generally the case in Europe, but rounded like domes, as in the Andes of South America. Their sides, however, are broken up by precipitous masses of tufa and conglomerate, intersected by deep ravines of the gloomiest character. They are covered with a thick shroud of ice and snow, but in their wombs seethe the fiery elements which ever and anon break forth into terrible activity. The eastern

AN ICELANDIC LANDSCAPE.

table-land and its mountain range is the most extensive, and contains Oërafa, the culminating point of Iceland. It is visible from a great distance at sea, like a white cloud suspended above the island. Its height is 6426 feet, and it springs from a vast mountain-mass; no fewer than 3000 square miles being perpetually burdened with ice and snow, at an altitude varying from 3000 to 6000 feet.

A very considerable portion of the island is occupied by the large glaciers which descend from the mountains, like frozen torrents, pushing forward into the lowlands, and even to the margin of the sea. These act as almost impassable barriers to communication between the various inhabited districts.

We have spoken of the two ranges of table-lands as about ninety to one hundred miles apart. The interspace forms a low broad valley, which opens at either extremity on the sea—an awful waste, a region of desolation, where man is utterly powerless; where the elements of fire and frost maintain a perpetual antagonism; where blade of grass is never seen, nor drop of water; where bird never wings its way, and no sign of life can be detected. It seems a realization of Dante's "circle of ice" in the "Inferno." The surface consists of lava streams, fissured by innumerable crevices; of rocks piled on rocks; of dreary glaciers, relieved by low volcanic cones. It is supposed that some remote portions of the inaccessible interior are less barren, because herds of reindeer have been seen feeding on the Iceland moss that fringes the borders of this dreary region. But there is no reason to believe that it can ever be inhabited by man.

MOUNT HEKLA, FROM THE VALLEY OF HEVITA.

The extremities of the valleys, where they approach the ocean, are the principal theatres of volcanic activity. At the northern end the best-known volcano is that of Hekla, which has attained a sinister repute from the terrific character of its eruptions. Of these six-and-twenty are recorded, the last having occurred in 1845–46. One lasted for six years, spreading devastation over a country which had formerly been the seat of a prosperous colony, burying the fields beneath a flood of lava, scoriæ, and ashes. During the eruption of September 2, 1845, to April 1846, three new craters were formed, from which columns of fire sprang to the height of 14,000 feet. The lava accumulated in formidable hills, and fragments of scoriæ and pumice-stone, weighing two hundredweight, were thrown to a distance of a league and a half; while the ice and snow which had lain on the mountain for centuries were liquefied, and poured down into the plains in devastating torrents.

. But the eruption of another of these terrible volcanoes, the Skaptá Jokul, which broke out on the 8th of May 1783, and lasted until August, was of a still more awful character. At that time the volcanic fire under Europe must have raged most violently, for a tremendous earthquake shattered a wide extent of Calabria in the same year, and a submarine volcano had flamed fiercely for many weeks in the ocean, thirty miles from the south-west cape of Iceland.

Its fires ceased suddenly; a series of earthquakes shook the island; and then Skaptá broke forth into sudden and destructive activity.

For months the sun was hidden by dense clouds of vapour, and clouds of volcanic dust were carried many hundreds of miles to sea, extending even to England and Holland. Sand and ashes, raised to an enormous height in the atmosphere, spread in all directions, and overwhelmed thousands of acres of fertile pasturage. The sulphurous exhalations blighted the grass of the field, and tainted the waters of river, lake, and sea, so that not only the herds and flocks perished, but the fish died in their poisoned element.

The quantity of matter ejected by the rent and shivered mountain was computed at fifty or sixty thousand millions of cubic yards. The molten lava flowed in a stream which in some places was twenty to thirty miles in breadth, and of enormous thickness; a seething, hissing torrent, which filled the beds of rivers, poured into the sea nearly fifty miles from its points of eruption, and destroyed the fishing on the coast. Some of the island-rivers were heated, it is said, to ebullition; others were dried up; the condensed vapour fell in whirls of snow and storms of rain. But dreadful as was the eruption itself, with its sublime but awful phenomena, far more dreadful were its consequences. The country within its range was one wide ghastly desert, a fire-blighted wilderness; and, partly from want of food, partly owing to the unwholesome condition of the atmosphere, no fewer than 9336 men,* 28,000 horses, 11,461 cattle, and 190,000 sheep, were swept away in the short space of two years. Even yet Iceland has scarcely recovered from the blow.

At the northern end of the great central valley the focus of igneous phenomena is found in a semicircle of volcanic heights which slope towards the eastern shore of the Lake Myvatr. Two of these are very formidable,—namely, Leirhnukr and Krabla on the north-east. After years of inaction, they suddenly broke out with tremendous fury, pouring such a quantity of lava into the Lake Myvatr, which measures twenty miles in circuit, that the water was in a state of ebullition for many days. On the sides of Mount Krabla, and at the base of this group of mountains, are situated various caldrons of boiling mineral pitch, the ruined craters of ancient volcanoes; and from their depths are thrown up jets of the molten matter, enveloped in clouds of steam, and accompanied by loud explosions at regular intervals.

But the most singular phenomena in this singular country, where frost and fire are continually disputing the pre-eminence, are the *Geysers*, or eruptive boiling springs. These all occur in the trachytic formation, are characterized by their high temperature, by holding siliceous matter in solution, which they deposit in the form of siliceous sinter, and by evolving large quantities of sulphuretted hydrogen gas.

Upwards of fifty geysers have been counted in the space of a few acres at the southern end

* A more moderate estimate says 1300 persons.

of the great valley. Some are constant, some periodical, some stagnant, some only slightly agitated. The grandest and most celebrated are the Great Geyser and Strokr, thirty-five miles north-west from Hekla. These, at regular intervals, hurl into the air immense columns of boiling water, to the height of one hundred feet, accompanied by clouds of steam and deafening noises. In the case of the Great Geyser, the jet issues from a shaft about seventy-five feet deep, and ten in diameter, which opens into the centre of a shallow basin, about one hundred and fifty feet in circumference. The basin is alternately emptied and filled: when filled, loud explosions are heard, the ground quivers, and the boiling water is forced upwards in gigantic columns. Thus the basin is emptied, and the explosions cease until it is refilled.

THE GREAT GEYSER.

Messrs. Descloiseaux and Bunsen, who, according to Mrs. Somerville, visited Iceland in 1846, found the temperature of the Great Geyser, at the depth of seventy-two feet, to equal 260° 30′ F. prior to a great eruption, reduced, after the eruption, to 251° 30′ F.; an interval of twenty-eight hours passing in silence.

About one hundred and forty yards distant is the Strokr (from *stroka*, to agitate), a circular well, forty-four feet deep, with a tube eight feet wide at its mouth, diminishing to little more than ten inches at a depth of twenty-seven feet. The surface of the water is in constant ebullition, while at the bottom the temperature exceeds that of boiling water by about twenty-four degrees. It appears, from experiments made by Donny, that water, long boiled, becomes more and more free from air, and that thus the cohesion of the particles is so much increased, that when the heat is sufficiently increased to overcome that cohesion, the production of steam is so considerable and so instantaneous as to induce an explosion. In this circumstance M. Donny finds an explanation of the phenomena of the Geysers, which are in constant ebullition for many hours, until, being almost purified from air, the intense internal or subterranean heat overcomes the cohesion of the particles, and thus an explosion takes place.

Lord Dufferin describes an eruption which he witnessed on the occasion of his visit to the Geysers, but for which he waited three days. Like pilgrims round some ancient shrine, he says, he and his friends kept patient watch; but the Great Geyser scarcely deigned to vouchsafe the slightest manifestation of its latent energies. Two or three times they heard a sound as of subterranean cannonading, and once an eruption to the height of about ten feet occurred. On the morning of the fourth day a cry from the guides made them start to their feet, and with one common impulse rush towards the basin. The usual underground thunder had already commenced. A violent agitation was disturbing the centre of the pool. Suddenly a dome of water lifted itself up to the height of eight or ten feet, then burst, and fell; immediately after which a shining liquid column, or rather a sheaf of columns, wreathed in robes of vapour, sprang into the air, and in a succession of jerking leaps, each higher than the last, flung their silver crests against the sky. For a few minutes the fountain held its own, then all at once appeared to lose its ascending power. The unstable waters faltered, drooped, fell, "like a broken purpose," back upon themselves, and were immediately sucked down into the recesses of their pipe.

The spectacle was one of great magnificence; but no description can give an accurate idea of its most striking features. The enormous wealth of water, its vitality, its hidden power, the illimitable breadth of sunlit vapour, rolling out in exhaustless profusion,—these combine to impress the spectator with an almost painful sense of the stupendous energy of nature's slightest movements.

The same traveller furnishes a very humorous account of the Strokr (or "churn").

It is, he says, an unfortunate Geyser, with so little command over its temper and stomach, that you can get a *rise* out of it whenever you like. Nothing more is necessary than to collect a quantity of sods and throw them down its funnel. As it has no basin to protect it from these liberties, you can approach to the very edge of the pipe, which is about five feet in diameter, and look down at the boiling water perpetually seething at the bottom. In a few minutes the dose of turf just administered begins to disagree with it; it works itself up into "an awful passion;" tormented by the qualms of incipient sickness, it groans and hisses, and boils up, and spits at you with malicious vehemence, until at last, with a roar of mingled pain and rage, it throws up into the air a column of water forty feet high, carrying with it all the sods that have been thrown in, and scattering them, scalded and half-digested, at your feet. So irritated has the poor thing's stomach become by the discipline it has undergone, that even long after all foreign matter has been thrown off, it continues retching and sputtering, until at last nature is exhausted, when, sobbing and sighing to itself, it sinks back into the bottom of its den.

The ground around the Geysers, for about a quarter of a mile, looks as if it had been "honeycombed by disease into numerous sores and orifices;" not a blade of grass grew on its hot, inflamed surface, which consisted of unwholesome-looking red livid clay, or crumpled shreds and shards of slough-like incrustations.

A region, corresponding in character to the desert mountain-mass we have been describing, stretches westward from it to the extremity of the ridge of the Snaefield Syssel, and terminates in the remarkable cone of Snaefield Jokul.

The island coasts exhibit a singularly broken outline, and the deep lochs or fiords, like those of Norway, only less romantic, dip into the interior for many miles, and throw off numerous branches. These fiords are wild and gloomy; dark, still inlets, with precipices on either side, a thousand feet in height, and the silence unbroken, save by the occasional wash of the waters, or the scream of a solitary ocean-bird. Inland, however, they assume a gentler character: they end in long narrow valleys, watered by pleasant streams, and bright with pasture. In these bits of Arcadia the inhabitants have built their towns and villages.

In the valleys on the north coast, which are adorned by clumps of willow and juniper, the soil is comparatively fertile; but the most genial scenery is found on the east, where, in some places, the birch-trees reach a height of twenty feet, and are of sufficient size to be used in house-building. The fuel used by the Icelanders is the drift-wood which the Gulf Stream brings from Mexico, the Carolinas, Virginia, and the River St. Lawrence.

In the south of the island the mean temperature is about 39°; in the central districts, 36°; in the north it rarely rises above 32°, or freezing-point. Thunder-storms, though rare in high latitudes, are not uncommon in Iceland; a circumstance which is due, no doubt, to the atmospheric disturbances caused by the volcanic phenomena. Hurricanes are frequent, and the days are few when the island is free from sea-mists. At the northern end the sun is always above the horizon in the middle of summer, and under it in the middle of winter; but absolute darkness does not prevail.

One of the most interesting places in Iceland is Thingvalla, where of old the "Althing," or supreme parliament, was wont to hold its annual assemblies, under the "Logmathurman," or president of the republic.

It is nothing more than a broad plain on the bank of the River Oxerá, near the point where the swift waters, after forming a noble cascade, sweep into the Lake of Thingvalla. Only a plain; but the scenery around it is indescribably grand and solemn. On either side lies a barren plateau, above which rises a range of snowy mountains, and from the plateaus the plain is cut off by deep chasms,—that of Almanna Gja on the east, and the Hrafna Gja on the west. It measures eight miles in breadth, and its surface is covered by a network of innumerable fissures and crevices of great depth and breadth. At the foot of the plain lies a lake, about thirty miles in circumference, in the centre of which two small crater-islands, the result of some ancient eruption, are situated. The mountains on its south bank have a romantic aspect, and that their volcanic fires are not extinct is shown by the clouds of vapour evolved from the hot springs that pour down their rugged sides. The actual meeting-place of the Althing was an irregular oval area, about two hundred feet by fifty, almost entirely surrounded by a crevice so broad and deep as to be impassable, except where a narrow causeway connected it with the adjacent plain, and permitted access to its interior. At one other point, indeed, the encircling chasm is so narrow that it may possibly be cleared at a leap; and the story runs that one Flosi, when hotly pursued by his enemies, did in this way escape them; but as falling an inch short would mean sure death in the green waters below, the chasm may be regarded as a tolerably sure barrier against intruders.

The ancient capital of the island was Skalholt, where, in the eleventh century, was founded the first school; an episcopal seat; the birthplace of a long line of Norse worthies, Islief the chronicler, Gissur the linguist, and Finnur Johnson the historian. But its glories have passed away; its noble cathedral has ceased to exist; and three or four cottages alone perpetuate the name of the once flourishing city.

The present capital is Reikiavik, to which, in 1797, were transferred the united bishoprics of Stoolum and Skalholt. It consists of a collection of wooden sheds, one story high, rising here and there into a gable end of greater pretensions, extending along a tract of dreary lava, and flanked at either end by a suburb of turf huts. On every side of it stretches a dreary lava-plain, and the gloom of the scorched and ghastly landscape is unrelieved by tree or bush. The white mountains are too distant to serve as a background to the buildings, but before the door of each merchant's house, facing the sea, streams a bright little pennon; and as the traveller paces the silent streets, whose dust no carriage-wheel has ever desecrated, the rows of flower-pots peeping out of the windows, between white muslin curtains, at once convince him that, notwithstanding their unostentatious appearance, within each dwelling reign "the elegance and comfort of a woman-tended home."

The prosperity of Reikiavik is chiefly due to its excellent harbour, and to the fish-banks in its neighbourhood, which supply it with an important commercial staple. In the summer and early autumn it is much visited by tourists, who start from thence to admire the wonders of Hekla, Skapta, and the Geysers; but its busiest time is in July, when the annual fair draws thither a crowd of fisher-folk and peasants. From a distance of forty and fifty leagues they come, with long trains of pack-horses, their stock-fish slung loose across the animals' backs, and their other wares packed closely in boxes or bags of reindeer-skin.

The Icelander is honest, temperate, hospitable, possessed with a fervent spirit of patriotism, and strongly wedded to the ancient usages. He is also industrious; and though his industry

HARBOUR OF REIKIAVIK.

is but scantily remunerated, he earns enough to satisfy his simple tastes. In the interior his chief dependence is on his cattle; and as grass is the main produce of his farm, his anxiety during the haymaking season is extreme. A bad crop would be almost ruin. He is, however, wofully ignorant of agriculture as a science, does but little for the improvement of the soil, and employs implements of the most primitive character. The process of haymaking in Iceland is thus described:—

The best crops are gathered from the "tún," a kind of home-park or paddock, comprising the lands contiguous to the farmstead—the only portion of his demesne to which the owner gives any special attention, and on the improvement of which he bestows any labour. This "tún" is enclosed within a wall of stone or turf, and averages an extent of two or three acres, though sometimes it reaches to ten. Its surface is usually a series of closely-packed mounds, like an overcrowded graveyard, with channels or water-runs between, about two feet deep.

Hither every person employed on the farm, or whom the farmer can engage, resorts, with short-bladed scythe and rake, and proceeds to cut down the coarse thick grass, and rake it up into little heaps.

Afterwards the mowers hasten to clear the neighbouring hill-sides and undrained marshes.

This primitive haymaking, so unlike the systematic operation which bears that name in England, is carried on throughout the twenty-four hours of the long summer day. The hay, when sufficiently dried, is made up into bundles, and tied with cords and thongs, and packed on the back of ponies, which carry it to the clay-built stalls or sheds prepared for it. And a curious sight it is to see a long string of hay-laden ponies returning home. Each pony's halter is made fast to the tail of his predecessor; and the little animals are so overshadowed and overwhelmed by their burdens, that their hoofs and the connecting ropes alone are visible, and they seem like so many animated haycocks, feeling themselves sufficiently made up, and leisurely betaking themselves to their resting-places.

During the protracted winter the Icelander, of course, can attend to no out-of-door labour, and passes his time within his hut, which, in many parts of the island, is not much superior to an Irish "cabin."

The lower part is built of rough stones up to a height of four feet, and between each course a layer of turf is placed, which serves instead of mortar, and helps to keep out the cold. The roof, made of any available wood, is covered with turf and sods. On the southern side the building is ornamented with doors and gable-ends, each of which is crowned by a weathercock. These doors are the entrances to the dwelling-rooms and various offices, such as the cow-shed, store-house, and smithy. The dwelling-rooms are connected by a long, dark, narrow passage, and are separated from each other by strong walls of turf. As each apartment has its own roof, the building is, in effect, an aggregate of several low huts, which receive their light through small windows in the front, or holes in the roof, covered with a piece of glass or skin. The floors are of stamped earth; the fireplace is made of a few stones, rudely packed together, while the smoke escapes through a hole in the roof, or through a cask or barrel, with the ends knocked out, which acts as chimney.

In some parts of the island lava is used instead of stones, and instead of wood the rafters are made of the ribs of whales. A horse's skull is the best seat provided for a visitor. Too often the same room serves as the dining, sitting, and sleeping place for the whole family, and the beds are merely boxes filled with feathers or sea-weed. There are, however, a few houses of a superior character, in which the arrangements are not much unlike those of a good old-fashioned English farm-house; the walls being wainscotted with deal, and the doors and staircase of the same material. A few prints and photographs, some book-shelves, one or two little pictures, decorate the sitting-room, and a neat iron stove, and massive chests of drawers, furnish it sufficiently.

From the houses we turn to the churches. In Reikiavik the church is a stone building, the only stone building in the town; but this is exceptional: most of the churches are not much better than the houses. We will be content, therefore, with a visit to the Reikiavik sanctuary, which is a neat and unpretending erection, capable of accommodating three or four hundred persons. The Icelanders are not opposed to a "decent ritual," and the Lutheran minister wears a black gown with a ruff round his neck. The majority of the congregation, here as everywhere else, consists of women; some few dressed in bonnets, and the rest wearing the national black silk skull-cap, set jauntily on one side of the head, with a long black tassel drooping to the shoulder, or else a quaint mitre-like structure of white linen, almost as imposing as the head-dress of a Normandy peasant. The remainder of an Icelandic lady's costume, we may add, consists of a black bodice, fastened in front with silver clasps, over which is drawn a cloth jacket, gay with innumerable silver buttons; round the neck goes a stiff ruff of velvet, embroidered with silver lace; and a silver belt, often beautifully chased, binds the long dark wadmal petticoat round the waist. Sometimes the ornaments are of gold, instead of silver, and very costly.

Towards the end of the Lutheran service, the preacher descends from the pulpit, and attiring himself in a splendid crimson velvet cope, turns his back to the congregation, and chants some Latin sentences.

Though still retaining in their ceremonies, says a recent traveller, a few vestiges of the old religion, though altars, candles, pictures, and crucifixes yet remain in many of their churches, the Icelanders are stanch Protestants, and singularly devout, innocent, and pure-hearted. Crime, theft, debauchery, cruelty are unknown amongst them; they have neither prison, gallows, soldiers, nor police; and in the manner of their lives mingles something of a patriarchal simplicity, that reminds one of the Old World princes, of whom it has been said, that they were "upright and perfect, eschewing evil, and in their hearts no guile."

In the rural districts, if such a phrase can properly be applied to any part of Iceland, the church is scarcely distinguishable from any other building, except by the cross planted on its roof. It measures, generally, from eight to ten feet in width, and from eighteen to twenty-four feet in length; but of this space about eight feet are devoted to the altar, which is divided off by a wooden partition stretching across the church, just behind the pulpit. The communion-table is nothing more than a small wooden chest or cupboard, placed at the end of the building, between two small square windows, each formed by a single common-sized pane of glass. Over the table is suspended a sorry daub, on wood, intended to represent the Last Supper. The walls, which are wainscotted, are about six feet high; and stout beams of wood stretch from side to side. On these are carelessly scattered a number of old Bibles, psalters, and loose leaves of soiled and antiquated manuscripts. The interior of the roof, the rafters of which rest on the walls, is also lined with wood. Accommodation, in the shape of a few rough benches, is provided for a congregation of thirty or forty souls.

Poor as are the churches, the pastors are still poorer. The best benefice in the island is worth not much more than £40 per annum; the average value is £10. The bishop himself does not receive more than £200. The principal support of the clergy, therefore, is derived from their globe-land, eked out by the small fees charged for baptisms, marriages, and funerals.

Such being the case, the reader will not be surprised to learn that the Icelandic clergy live miserably and work hard. They assist in the haymaking; they hire themselves out as herdsmen; they act as the leaders of the caravans of pack-horses which carry the produce of the island to the ports, and return loaded with domestic necessaries; and they distinguish themselves as blacksmiths, as veterinarians, and shoers of horses.

Dr. Henderson gives an interesting and graphic account of a visit he paid to one of these "poor parsons," John Thorlukson, who, while supporting himself by drudgery of this painful kind, translated Milton's "Paradise Lost" and Pope's "Essay on Man" into Icelandic.

"Like most of his brethren, at this season of the year," says Dr. Henderson, "we found him in the meadow assisting his people at haymaking. On hearing of our arrival he made all the haste home which his age and infirmity would allow, and bidding us welcome to his lowly abode, ushered us into the humble apartment where he translated my countrymen into Icelandic. The door was not quite four feet in height, and the room might be about eight feet in length by six in breadth. At the inner end was the poet's bed; and close to the door, over against a small window, not exceeding two feet square, was a table where he committed to paper the effusions of his muse. On my telling him that my countrymen would not have forgiven me, nor could I have forgiven myself, had I passed through this part of the island without paying him a visit, he replied that the translation of Milton had yielded him many a pleasant hour, and often given him occasion to think of England."

It is true that this passage was written some fifty-five years ago, but the condition of the clergy of Iceland has not much improved in the interval.

Travelling in Iceland, even under the more favourable conditions brought about by a constant influx of tourists, is not to be achieved without difficulty and discomfort. Not only is the country destitute, necessarily, of inns and the usual arrangements for the convenience of travellers, but much, very much, depends upon the weather. With a bright sky overhead, it is possible to regard as trivial and unworthy of notice the small *désagréments* which, in bad weather, develop into very serious annoyances. The only mode of travelling is on horseback, for as there are no roads, carriages would be useless; while the distances between the various points of interest are too great, the rivers too violent, and the swamps too extensive for pedestrian tours to be undertaken. Even the most moderate-minded tourist requires a couple of riding-horses for himself, a couple for his guide, and a couple of pack-horses; and when a larger company travels, it expands into a cavalcade of from twenty to thirty horses, tied head to tail, which slowly pick their way over rugged lava-beds or dangerous boggy ground.

It is one thing, as Lord Dufferin remarks, to ride forty miles a day through the most singular scenery in the world, when a glorious sun brings out every feature of the landscape into startling distinctness, transmuting the dull tormented earth into towers, domes, and pinnacles of shining metal, and clothes each peak in a robe of many-coloured light, such as the "Debatable Mountains" must have been in Bunyan's dream; and another to plod over the same forty miles, wet to the skin, seeing nothing but the dim gray bases of the hills, which rise you know not how, and care not where. "If, in addition to this, you have to wait, as very often must be the case, for many hours after your own arrival, wet, tired, hungry, until the baggage-train, with the tents and food, shall have come up, with no alternative in the meantime but to lie shivering inside a grass-roofed house, or to share the quarters of some farmer's family, whose domestic arrangements resemble in every particular those which Macaulay describes as prevailing among the Scottish Highlanders a hundred years ago; and if, finally, after vainly waiting for some days to see an eruption which never takes place, you journey back to Reikiavik under the same melancholy conditions, it will not be unnatural that, on returning to your native land, you should proclaim Iceland, with her geysers, to be a sham, a delusion, and a snare!"

There are no bridges in Iceland; no bridges, except, indeed, a few planks flung across the Bruera, and a swing-bridge, or kláfe, which spans the Jokülsa; and, as is still the case in some parts of the Scottish Highlands, the traveller must ford the streams, which are always rapid, and sometimes inconveniently deep. The passage of a river is, therefore, a formidable enterprise, as may be inferred from the experiences of Mr. Holland and other travellers.

The guide leads the way, and the caravan follow obediently in his wake, stemming, as best they can, the swift impetuous torrent. Often the boiling water rises high against the horse's shoulders, and dashes clouds of spray in the face of the riders. The stream is so furiously fast that it is impossible to follow the individual waves as they sweep by, and to look down at it almost makes one dizzy. Now, if ever, is the time for a firm hand, a sure seat, and a steady eye: not only is the current strong, but its bed is full of large stones, which the horse cannot see through the dark waters; and should he fall, the torrent will carry you down to the sea, whose white breakers are plainly visible as they crawl along the resounding beach at a mile's distance. Happily, though hungry for prey, they will not be satisfied. Swimming would be of no use, but an "Icelandic water-horse" seldom blunders or makes a false step. But another danger lies in the masses of ice swept down by the whirling waves, many of which are sufficiently large to topple over horse and rider.

How the horses are able to stand against such a stream is every traveller's wonder; nor would they do so unless they were inured to the enterprise from their very youth. The Icelanders who live in the interior keep horses known for their qualities in fording difficult rivers, and never venture to cross a dangerous stream unless mounted on an experienced "water-horse."

The action of the Icelandic horses in crossing a swift river is very peculiar. They lean all their weight *against* the current, so as to oppose it as much as possible, and move onwards with a characteristic side-step. This motion is not agreeable. It feels as if your horse were marking time, like soldiers at drill, without gaining ground, and as the progress made is really very slow, the shore from which you started seems to recede from you, while that to which you are bound does not seem to draw nearer."

In the mid-stream the roar of the waters is frequently so great that the travellers cannot make their voices audible to one another. There is the swirl of the torrent, the seething of the spray, the crunching of the floating ice, the roll of stones and boulders against the bottom,—and all these sounds combine in one confused chaotic din. Up to this point, a diagonal line, rather down stream, is cautiously followed; but when the middle is reached, the horses' heads are turned slightly *towards* the current, and after much effort and many risks the opposite bank is reached in safety.

Lord Dufferin says, with much truth, that the traveller in Iceland is constantly reminded of the East. From the earliest ages the Icelanders have been a people dwelling in tents. In the days of the ancient Althing, the legislators, during the entire session, lay encamped in movable booths around the place of council. There is something patriarchal in their domestic polity, and the very migration of their ancestors from Norway was a protest against the antagonistic principle of feudalism. No Arab could be prouder of his high-mettled steed than the Icelander of his little stalwart, sure-footed pony: no Oriental could pay greater attention to the duties of hospitality; while the solemn salutation exchanged between two companies of travellers, as they pass each other in what is universally called "the desert," is not unworthy of the stately courtesy of the gravest of Arabian sheikhs.

It is difficult to imagine anything more multifarious than the cargo which these caravans import into the inland districts: deal boards, rope, kegs of brandy, sacks of rye or wheaten flour, salt, soap, sugar, snuff, tobacco, coffee; everything, in truth, which is necessary for domestic consumption during the dreary winter season. In exchange for these commodities the Icelanders give raw wool, knitted stockings, mittens, cured cod, fish-oil, whale-blubber, fox-skins, eider-down, feathers, and Iceland moss. The exports of the island in wool amount to upwards of 1,200,000 lbs. of wool yearly, and 500,000 pairs of stockings and mittens.

ICELANDERS FISHING FOR NARWHAL.

Iceland offers abundant sport to the enthusiast in fishing. The streams are well supplied with salmon; while the neighbouring seas abound in seals, torsk, and herrings. The narwhal-fishery is also carried on, and has its strange and exciting features. The implement used is simply a three-pronged harpoon, like a trident, with which the fisherman strikes at the fish as they rise to the surface; and his dexterity and coolness are so great that he seldom misses his aim.

Numerous works, in English, have been written upon Iceland and the Icelanders; the most trustworthy are those by Dr. Henderson, Professor Forbes, Holland, Chambers, and Lord Dufferin. The King of Denmark visited Iceland in 1874.

CHAPTER VII.

THE ESKIMOS.

THE land of the Eskimos is of very wide extent. From Greenland and Labrador they range over all the coasts of Arctic America to the extreme north-eastern point of Asia. Several of the Eskimo tribes are independent; others acknowledge the rule of Great Britain, Denmark, Russia, and more recently of the United States. The whaler meets with them on the shores of Baffin Bay, and in the icy sea beyond Behring Straits; the explorer has tracked them as far as Smith Sound, the highway to the North Pole; and while they descend as low as the latitude of Vienna, they rove as far north as the 81st and 82nd parallels. They are the aborigines of the deserts of ice and snow, the ancient masters of the Arctic wilderness, and all Polar America is their long-acknowledged domain. To a certain extent they are nomadic in their habits; compelled to migrate by the conditions of the climate in which they live, and forced to seek their scanty sustenance in a new locality when they have exhausted the capabilities of any chosen habitat. As Mr. Markham tells us, traces of former inhabitants are found throughout the gloomiest wastes of the Arctic regions, in sterile and silent tracts where now only solitude prevails. These wilds, it is known, have been uninhabited for centuries; yet they are covered with memorials of wanderers or of sojourners of a bygone age. Here and there, in Greenland, in Boothia, on the American coast, where life is possible, the descendants of former nomads are still to be found.

Arctic discovery, as yet, has stopped short at about 82° on the west coast, and 76° on the east, of Greenland. These two points are about six hundred miles apart. There have been inhabitants at both points, though they are separated by an uninhabitable interval from the settlements further south; we may conclude, then, that the *terra incognita* further north is also or has been inhabited. In 1818 it was discovered that a small tribe of Eskimos inhabited the bleak west coast of Greenland between 76° and 79° N. They could not penetrate to the south on account of the glaciers of Melville Bay; they could not penetrate to the north, because all progress in that direction is forbidden by the great Humboldt glacier; while the huge interior glacier of the *Sernik-sook* pent them in upon the narrow belt of the sea-coast. These so-called "Arctic Highlanders" number about one hundred and forty souls, and throughout the winter their precarious livelihood depends on the fish they catch in the open pools and water-ways. Under similar conditions, it is probable that Eskimo tribes may be existing still further north; or if, as geographers suppose, an open sea really surrounds the Pole, and a warmer atmosphere prevails, the conditions of their existence will necessarily be more favourable.

Before we come to speak of the characteristics of the Eskimos, we must briefly notice the Danish settlements in Greenland, which are gradually attracting no inconsiderable number of them within the bounds of civilization. These are dotted along the coast, like so many centres of light and life; but the most important, from a commercial point of view, are Upernavik, Jacobshav'n, and Godhav'n.

Upernavik is the chief town of a district which extends from the 70th to the 74th degree of north latitude, and enjoys the distinction of being the most northerly civilized region in the world. Its northern boundary represents the furthest advance of civilization in its long warfare against the Arctic climate.

The town of Upernavik is situated on the summit of a mossy hill which slopes to the head of a small but sheltered harbour. It contains a government-house, plastered with pitch and tar; a shop or two; lodging-houses for the Danish officials; some timber huts, inhabited by Danes;

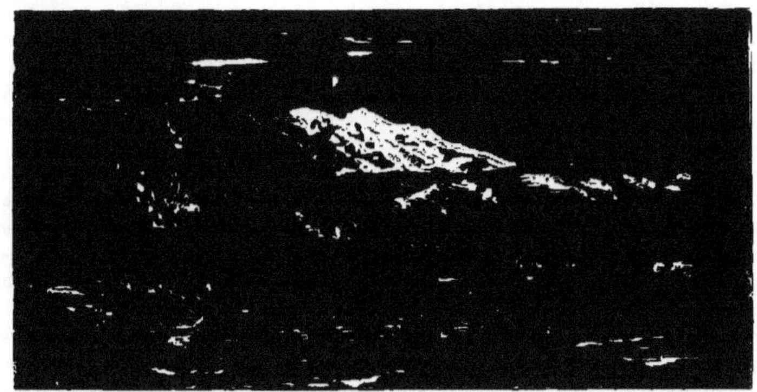

UPERNAVIK, GREENLAND.

and a number of huts of stone and turf, intermingled with seal-skin tents, which accommodate the natives. Its principal evidences of civilization are its neat little church and parsonage.

The inhabitants are chiefly occupied in fishing and hunting, and in the manufacture of suitable clothing for the protection of the human frame against the winter cold. Reindeer, seal, and dog skins are deftly converted into hoods, jackets, trousers, and boots. The last-named are triumphs of ingenuity. They are made of seal-skin, which has been tanned by alternate freezing and thawing; are sewed with sinew, and "crimped" and fitted to the foot with equal taste and skill. Dr. Hayes informs us that the Greenland women, not exempt from the love of finery characteristic of their sex, trim their own boots in a perfectly bewitching manner, and adopt the gayest of colours. Red boots, or white, trimmed with red, he says, seemed most generally worn, though there was no more limit to the variety than to the capriciousness of the fancy which suggested it. And it would be difficult to imagine a more grotesque spectacle than is presented by the crowd of red, and yellow, and white, and purple, and blue-legged women who crowd the beach whenever a strange ship enters the harbour.

The population of Upernavik numbers now about two hundred and fifty souls; comprising some forty or fifty Danes, a larger number of half-breeds, the remainder being native Greenlanders,—that is, Eskimos.

DISCO ISLAND, GREENLAND.

In describing *one* Danish settlement we describe all, for they present exactly the same characteristics, the difference between them being only a question of population.

Jacobshav'n and Godhav'n are situated on the island of Disco, which is separated from the

GODHAV'N, DISCO ISLAND, GREENLAND.

west coast of Greenland by Weygat Strait, and has been described as one of the most remarkable localities in the Arctic World. The tradition runs that it was translated from a southern region to its present position by a potent sorcerer; and an enormous hole in the rock is pointed out as the gully through which he passed his rope. It is a lofty island, and its coast is belted round by high trap cliffs, of the most imposing aspect. Near its south-west extremity, in lat. 69° S., a low rugged spur or tongue of granite projects into the sea for about a mile and a half,—a peninsula at low water, and an island at high water,—and forms the snug little recess of Godhav'n, or Good Harbour. To the north of the bay, in face of rocky cliffs, which rise perpendicularly from the sea to a height of 2000 feet, lies the town of the same name, which our English whalers know as Lievely, probably a corruption of the adjective *lively;* for the tiny colony is the metropolis of Northern Greenland; and since the beginning of the present century has been the favourite rendezvous of the fishing fleets and expeditions of discovery.

Further to the north lies Jacobshav'n, which possesses a celebrity of its own as one of the most ancient of the Moravian mission-stations in the north of Greenland. Besides a church, it

DANISH SETTLEMENT OF JACOBSHAV'N, GREENLAND.

boasts of a college for the education and training of natives who desire to be of service to their fellow-countrymen in the capacity of catechists or teachers. So great has been the industry, and so well deserved is the influence of the missionaries, that it is difficult now to find an Eskimo woman in this part of Greenland who cannot read and write. Prior to the Danish colonisation of Greenland, the language of the natives was exclusively oral. Only through the medium of speech could they represent their simplest ideas; and the picture-writing of the North American Indians was beyond their skill. But the missionaries have raised the Eskimo tongue into the rank of written languages. At Godthaab a printing-press is in full operation, and has already produced some very interesting historical narratives and Eskimo traditions.

As is the case with all the Greenland colonies, Jacobshav'n owes its prosperity to the seal-fishing. Moreover, the Greenland, or "right" whale, in its annual migrations southward, enters the neighbouring waters during the month of September, and furnishes employment to the fishing population.

In the neighbourhood of Jacobshav'n an enormous glacier, one of the offshoots of the great central *mer de glace* of Greenland, finds its way to the sea. Yet the temperature is said to be milder than at Godhav'n.

The following remarks apply, of course, to those Eskimos who still lead a nomadic life, and have profited little or nothing by the Christian civilization of the Danish settlements and Moravian missions.

Among themselves the Eskimos are known as *Inuits*, or "men;" the seamen of the Hudson Bay ships have long been accustomed to call them *Seymos* or *Suckemos*—names derived from the cries of *Seymo* or *Teymo* with which they hail the arrival of the traders; while the old Norsemen designated them, in allusion to their discordant shouts, or by way of expressing their infinite contempt, *Skrælingers*, "screamers" or "wretches."

The European feels impelled to pity the hard fate which condemns them to inhabit one of the dreariest and most inhospitable regions of the globe, where only a few mosses and lichens, or plants scarcely higher in the scale of creation, can maintain a struggling existence; where land animals and birds are few in number; and where human life would be impossible but for the provision which the ocean waters so abundantly supply. As they live in a great degree upon fish and the cetaceans, they dwell almost always near the coast, and never penetrate inland to any considerable distance.

In the east the Eskimos, for several centuries, have been subjected to the civilizing influences of the English and the Dutch; in the west, they have long been under the iron rule of the Muscovite. In the north and the centre their intercourse with Europeans has always been casual and inconsiderable. It will therefore be understood that the different branches of this wide-spread race must necessarily exhibit some diversity of character, and that the same description of manners and mode of life will not in all points apply with equal accuracy to the savage and heathen Eskimos of the extreme northern shores and islands, the Greek Catholic Aleüts, the faithful servants of the Hudson Bay Company, and the disciples of the Moravian Brethren in Labrador or Greenland. Yet the differences are by no means important, and it may be doubted whether any other race, living under such peculiar conditions, and extending over so vast an area, can show so few and such inconsiderable specific varieties. When one thinks of an Eskimo, one naturally calls up a certain image to one's mind: that of a man of moderate stature or under medium size, with a broad flat face, narrow tapering forehead, and narrow or more or less oblique eyes; and this image or type will be found to be realized throughout the length and breadth of Eskimo America. The Eskimo, generally speaking, would seem to have sprung from a Mongol stock; at all events, he can claim no kinship with the Red Indians. Happily for Europeans, if inferior to the latter in physical qualities, he is superior in generosity and amiability of disposition.

The Eskimos are sometimes spoken of as if they were dwarfs or Lilliputians, but such is not the case. They are shorter than the average Frenchman or Englishman, but individuals measuring from five feet ten inches to six feet have been found in Camden Bay. Dr. Kane speaks of Eskimos in Smith Strait who were fully a foot taller than himself. It is true of the females, however, that they are comparatively little.

The Eskimos are a stalwart, broad-shouldered race, considerably stronger than any other of the races of North America. In both sexes the hands and feet are small and well-shaped.

Their muscles are strongly developed, owing to constant exercise in hunting the seal and the walrus. They are also powerful wrestlers, and on no unequal terms could compete with the athletic celebrities of Devon and Cornwall. Their physiognomy, notwithstanding its lack of beauty, is far from displeasing; its expression is cheerful and good-tempered, and the long winter night does not seem to sadden their spirits or oppress their energies. The females are well made, and though not handsome, are scarcely to be stigmatized as ugly. Their teeth are very white and regular; and their complexion is warm, clear, and good. It is true that it cannot be seen to advantage, owing to the layers of dirt by which it is obscured; but it is not much darker than a dark brunette, and as for the dirt—well, perhaps, it is preferable to cosmetics!

Even in the Arctic World, woman seems conscious of the influence of her charms, and man seems willing to recognize it. They plait their black and glossy hair—these Eskimo beauties!—with much care and taste; and they tattoo their forehead, cheeks, and chin with a few curved lines, which produce a not altogether unpleasant effect.

From Behring Straits eastward, as far as the river Mackenzie, the males pierce the lower lip near each angle of the mouth, in order to suspend to it ornaments of blue or green quartz, or of ivory, shaped like buttons. Some insert a small ivory quill or dentalium shell in the cartilage of the nose. They decorate themselves, moreover, with strings of glass beads; or when and where these cannot be obtained, with strings of the teeth of the musk-ox, wolf, or fox; hanging them to the tail of the jacket, or twining them round the waist like a girdle.

The influence of climate upon dress is a subject which we commend to the notice of art-critics and æsthetic philosophers. Within the Arctic Circle the problem to be solved is, how to obtain the greatest amount of protection for the person, without rendering the costume too heavy or cumbrous; and the Eskimos have succeeded in solving it satisfactorily. They can defy the rigour of the Arctic winter, its extreme cold, its severest gales, and pursue their avocations in the open air even in the dreariness of the early winter twilight, so cleverly adapted is their garb to the conditions under which they live. Their boots, made of seal-skin, and lined with the downy skins of birds, are thoroughly waterproof; their gloves are large, but defend the hands from frost-bite: they wear two pair of breeches, made of reindeer or seal-skin, of which the under pair has the close, warm, stimulating hair close to the flesh; and two jackets, of which the upper one is provided with a large hood, completely enveloping the head and face, all but the eyes. The women are similarly attired, except that their outer jacket is a little longer, and the hood, in which they carry their children, considerably larger; and that, in summer, they substitute for the skin-jacket a water-tight shirt, or *kamleika*, made of the entrails of the seal or walrus. They sew their boots so tightly as to render them impervious to moisture, and so neatly that they may almost be included in the category of works of art. In Labrador the women carry their infants in their boots, which have a long pointed flap in front for the purpose.

In a preceding chapter we have spoken incidentally of the Eskimo huts. These, like the Eskimo dress, are admirably adapted to the circumstances of the country and the nature of the climate. The materials used are either frozen snow, earth, stones, or drift-wood. The snow-hut is a dome-shaped edifice, constructed in the following manner:—

First, the builders trace a circle on the smooth level surface of the snow, and the snow gathered within the area thus defined is cut into slabs, and used for building the walls, leaving the ice underneath to serve as the flooring.

The crevices between the slabs, and any accidental fissures, are closed up by throwing a few shovelfuls of loose snow over the building. Two men are generally engaged in the work; and when the dome is completed, the one within cuts a low door, through which he creeps. As the walls are not more than three or four inches thick, they admit a soft subdued light into the interior, but a window of transparent ice is generally added. Not only the hut, but the furniture inside it, is made of snow; snow seats, snow tables, snow couches—the latter rendered comfortable by coverings of skins. To exclude the cold outer air, the entrance is protected by an ante-chamber and a porch; and for the purposes of intercommunication, covered passages are carried from one hut to another.

The rapidity with which these snow-huts are raised is quite surprising, and certainly affords a vivid illustration of the old saying that "practice brings perfection." Captain M'Clintock for

BUILDING AN ESKIMO HUT.

a few nails hired four Eskimos to erect a hut for his ship's crew; and though it was twenty-four feet in circumference, and five and a half feet in height, it was erected in a single day.

Much ingenuity is frequently displayed in their construction.

Dr. Scoresby, in 1824, found some deserted huts on the east coast of Greenland, which showed no little constructive skill on the part of their builders.

A horizontal tunnel, about fifteen feet in length, and so low that a person entering it was compelled to crawl on his hands and knees, opened with one end to the south, while the other end terminated in the interior of the hut. This rose but slightly above the surface of the earth, and being generally overgrown with moss or grass, could scarcely be distinguished from the neighbouring soil. It resembled, indeed, a large ant-hill, or the work of a mammoth mole! In some cases the floor of the tunnel was on a level with that of the hut; but more frequently it slanted downwards and upwards, so that the colder, and consequently heavier, atmospheric air was still more completely prevented from mixing too quickly with the warmer air within. The other arrangements exhibited the same ingenuity in providing against the inconveniences of a rigorous climate.

From the huts of the Eskimos we pass to their boats.

The *kayak* or *baidar* is as good in its way as the light and swift canoe of the Polynesian islanders. It consists of a narrow, long, and light wooden framework, covered water-tight with seal-skin, with a central aperture for the body of the rower. Sometimes the frame is made of seal or walrus bone. The Eskimo takes his seat in his buoyant craft, with legs outstretched, and binds a sack—which is made from the intestines of the whale, or the skins of young seals—so tightly round his waist, that even in a rolling sea the boat remains water-tight. Dexterously and rapidly using his paddle, with his spear or harpoon before him, and preserving his equilibrium with marvellous steadiness, he darts over the waves like an arrow; and even if upset,

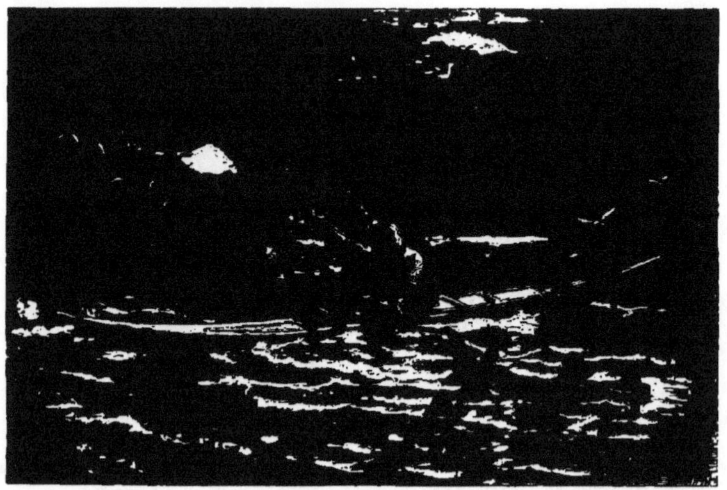

THE ESKIMO KAYAK.

speedily rights himself and his buoyant skiff. The *oomiak*, or woman's boat, consists in like manner of a framework covered with seal-skins; but it is large enough to accommodate ten or twelve people, with benches for the women who row or paddle. The mast supports a triangular sail, made of the entrails of seals, and easily distended by the wind.

It has been observed that a similar degree of inventive and executive skill is displayed by the Eskimos in their spears and harpoons, their fishing and hunting implements. Their oars are tastefully inlaid with walrus teeth; they have several kinds of spears or darts, according to the character of the animal they intend to hunt; and their bows, with strings of seal-gut, are so strong and elastic as to drive a six-foot arrow a really considerable distance. The harpoons and spears used in killing whales or seals have long shafts of wood or bone, and the barbed point is so constructed that, when lodged in the body of an animal, it remains imbedded, while the shaft attached to it by a string is loosened from the socket, and acts as a buoy. Seal-skins filled with air, like bladders, are also employed as buoys for the whale-spears, being stripped from the animal with such address that all the natural apertures are easily made air-tight.

Fish-hooks, knives, and spear or harpoon heads, the Eskimos make of the horns and bones of the deer. In constructing their sledges, and roofing their huts, they have recourse to the ribs of the whale, when drift-wood is not available. Strips of seal-skin hide are a capital substitute for cordage, and cords for nets and bow-strings are manipulated from the sinews of musk-oxen and deer.

THE ESKIMO OOMIAK.

A strange and deadly antagonism prevails between the Eskimos and the Red Indians. On the part of the latter it would seem to originate in jealousy, for the Eskimos are superior in skill, social habits, general intelligence, personal courage, and strength; on the part of the latter, in the necessity for self-defence and the provocations they have received from a sanguinary enemy.

Hence, the Indians inhabiting the borders of the Polar World seek every opportunity of surprising and massacring the inoffensive Eskimos. Hearne relates that, in the course of his expedition to the Coppermine River, the Indians who accompanied him obtained information that a party of Eskimos had raised their summer huts near the river-mouth. In spite of his generous efforts, they resolved on destroying the peaceful settlement. Stealthily they made their approach, and when the midnight sun touched the horizon, they swooped down, with a frightful yell, on their unfortunate victims, not one of whom escaped. With that love of torture which seems inherent in the Red Indian, they did their utmost to intensify and prolong the agonies of the sufferers; and one aged woman had both her eyes torn out before she received her death-blow. The scene where this cruel slaughter took place is known to this day as the "Bloody Falls."

Dr. Kane supplies some interesting particulars of a party of Eskimos with whom he became acquainted during his memorable expedition. The intimacy began under unfavourable circumstances, for three of the party had been detected in a scandalous theft, had attempted to carry off their plunder, were pursued, overtaken, and punished. Soon afterwards, Metek, the head man or chief, arrived on the scene, and a treaty of peace was concluded.

On the part of the *Inuit*, or Eskimos, it ran as follows:—

"We promise that we will not steal. We promise we will bring you fresh meat. We promise we will sell or lend you dogs. We will keep you company whenever you want us, and show you where to find the game."

On the part of the *Kablunah*, or white men, it ran as follows:—

"We promise that we will not visit you with death or sorcery, nor do you any hurt or mischief whatsoever. We will shoot for you on our hunts. You shall be made welcome aboard ship. We will give you presents of needles, pins, two kinds of knife, a hoop, three bits of hard wood, some fat, an awl, and some sewing-thread; and we will trade with you of these and everything else you want for walrus and seal meat of the first quality."

The treaty, says Dr. Kane, was not solemnized by an oath; but it was never broken.

The Eskimo settlement at Anatoak, lat. 73° N, on the shore of Smith Strait, near Cape Inglefield, seems to merit description.

The hut or igloë was a single rude elliptical apartment, built not unskilfully of stone, the outside lined with sods. At its further end, a rude platform, also of stone, was raised about a foot above the entering floor. The roof was irregularly curved. It was composed of flat stones, remarkably large and heavy, arranged so as to overlap each other, but apparently without any intelligent application of the principle of the arch. The height of this cave-like abode barely permitted one to sit upright. Its length was eight feet, its breadth seven feet, and an expansion of the tunnelled entrance made an appendage of perhaps two feet more.

The true winter-entrance is called the *tossut*. It is a walled tunnel, ten feet long, and so narrow that a man can hardly crawl along it. It opens outside below the level of the igloë, into which it leads by a gradual ascent.

Thus the reader will see that the hut at Anatoak was constructed on the same principles as the huts discovered by Dr. Scoresby.

Time had done its work, says Dr. Kane, on the igloë of Anatoak, as among the palatial structures of more southern deserts. The entire front of the dome had fallen in, closing up the *tossut*, or tunnel, and forcing visitors and residents to enter at the solitary window above it. The breach was wide enough to admit a sledge-team; but the Eskimos showed no anxiety to close it up. Their clothes saturated with the freezing water of the floes, these men of iron gathered round a fire of hissing and flaring whale's blubber, and steamed away in apparent comfort. The only departure from their usual routine was suggested probably by the open roof and the bleakness of the night; and therefore they refrained from stripping themselves naked before coming into the hut, and hanging up their dripping vestments to dry, like a votive offering to the god of the sea.

Their kitchen implements were remarkable for simplicity. "A rude saucer-shaped cup of seal-skin, to gather and hold water in, was the solitary utensil that could be dignified as table-furniture. A flat stone, a fixture of the hut, supported by other stones just above the shoulder-blade of a walrus,—the stone slightly inclined, the cavity of the bone large enough to hold a moss-wick and some blubber; a square block of snow was placed on the stone, and, as the hot smoke circled round it, the seal-skin saucer caught the water that dripped from the edge. They had no vessel for boiling; what they did not eat raw they baked upon a hot stone. A solitary coil of walrus-line, fastened to a movable lance-head (*noon-ghak*), with the well-worn and well-soaked clothes on their backs, completed the inventory of their effects."

The Eskimos entertained Dr. Kane and his companions with a choral performance, singing their rude, monotonous song of "Amna Ayah" till the unfortunate white men were almost maddened by the discord. They improvised, moreover, a special chant in their honour, which they repeated with great gravity of utterance, invariably concluding with the sonorous and complimentary refrain of "*Nalegak! nalegak! nalegak-soak!*"—"Captain! captain! great captain!" The chant ran as follows:—

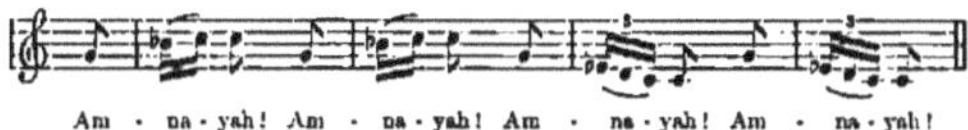
Am - na - yah! Am - na - yah! Am - na - yah! Am - na - yah!

In the early spring the Eskimos resume their hunting expeditions, and their snow-covered huts are transformed into scenes of the liveliest activity. Stacks of jointed meat, chiefly walrus, are piled upon the ice-foot; the women stretch the hide for sole-leather, and the men collect a store of harpoon-lines for the winter. Tusky walrus heads stare at the spectacle from the snow-bank, where they are stowed for their ivory; the dogs are tethered to the ice; and the children, each one armed with the curved rib of some big walrus or seal, play ball and bat among the snow-drifts.

The quantity of walrus meat which the Eskimos accumulate during a season of plenty should certainly raise them above all risk of winter want; but other causes than improvidence render their supplies scanty. They are never idle; they hunt incessantly without the loss of a day. When the storms prevent the use of the sledge, they occupy themselves in stowing away the spoils of previous hunts. For this purpose they dig a pit either on the mainland, or, which is preferred, on an island inaccessible to foxes, and the jointed meat is stacked inside, and covered with heavy stones.

The true explanation of the scarcity from which these people so frequently suffer is the excessive consumption in which they indulge during the summer season. By their ancient laws all share in common; and since they migrate in numbers when their necessities press them, the tax on each separate settlement is excessive. The quantity which the members of a family consume seems excessive to a stranger; yet it is not the result of inconsiderate gluttony, but due to their peculiarities of life and organization. In active exercise, and under the influence of exposure to a severe temperature, the waste of carbon must be enormous.

When in-doors, and at rest, engaged upon their ivory harness-rings, fowl-nets, or other household gear, they eat, as many eat in more civilized lands, for mere animal enjoyment, and to pass away the time. But when engaged in the chase, they take but one meal a day, and that not until the day's labour is ended. They go out upon the ice without breakfast, and seldom eat anything until their return. Dr. Kane estimates the average ration of an Eskimo in a season of plenty at eight to ten pounds of meat a day, with soup and water to the extent of half a gallon. Such an allowance might almost have satisfied the appetite of Gargantua!

Dr. Hayes, in the course of his adventurous Arctic boat-journey, held much intercourse with the Eskimos, and his impressions, on the whole, would seem to have been highly favourable.

His sketch of a couple whom he met in the neighbourhood of the Eskimo colony of Netlik is very amusing.

He describes them as a most *un*human-looking pair. Everything on and about them told of the battle they fought so gallantly and patiently with the elements. From head to foot they were invested in a coat of ice and snow. Shapeless lumps of whiteness, they resembled the snow-kings or statues which boys delight in making, except that they possessed the faculty of motion. Their long, heavy fox-skin coats, reaching nearly to the knees, and surmounted by a hood, covering, like a round lump, all of the head but the face, the bear-skin pantaloons and boots and mittens were saturated with snow. Their long, black hair, which fell from beneath their hoods over their eyes and cheeks, their eyelashes, the few hairs growing upon their chins, the rim of fur around their faces, all glittered with white frost—the frozen moisture of their breath. Each carried in his right hand a whip, and in his left a lump of frozen meat and blubber. The meat they flung down on the floor of Dr. Hayes' hut; then, without pausing for an invitation, they thrust their whipstocks under the rafters, and divesting themselves of their mittens and outer garments, hung them thereon. Underneath their frosty coats they wore a warm, close shirt of bird-skins.

In the same bold explorer's narrative of his voyage of discovery in 1860, two other Eskimos figure very conspicuously; and one of these, named Hans, would seem to have been a very fair type of the Eskimo character. Hans, we may observe, had originally served in Dr. Kane's expedition, and had then gained the confidence of Dr. Hayes; so that when the latter undertook his own memorable voyage, he became anxious to secure the Eskimo's services.

When his ship had crossed Melville Bay, and lay in the grim shadows of Cape York, Dr. Hayes bethought himself of the Eskimo hunter. He remembered to have heard that Hans had fallen in love, and taken a wife, and repaired, with her at his side, to share the fortunes of the wild Eskimos who inhabit the remote northern shores of Baffin Bay.

But Dr. Hayes felt confident that the hunter, having known something of the superior comfort and happiness of the social life of civilization, would soon weary of his voluntary banishment, and of the penury and hardships of the existence of the Eskimo nomads. He made up his mind that Hans would return to Cape York, and there take up his residence, in the hope of being picked up by some passing ship.

So Dr. Hayes stood close inshore, to find that his conjectures were completely realized. As he sailed along the coast he discovered a group of human beings eagerly endeavouring by signs and gestures to attract attention. Heaving the schooner to, he and his second in command, Mr.

13. HAYES FALLS IN WITH HANS THE HUNTER.

Sonntag, went ashore in a boat, and there was Hans! The Eskimo recognized both of them immediately, and called them by name.

We may adopt the remainder of Dr. Hayes' interesting little episode, because it illustrates the ingrained selfishness, or self-concentration, of the Eskimo character.

Hans had deteriorated greatly during his residence with the wild Eskimos, and he had sunk to their level of filthy ugliness. He was accompanied by his wife, who carried her first-born in a hood upon her back; his wife's brother, a quick-eyed boy of twelve years; and his wife's mother, "an ancient dame with voluble and flippant tongue." They were all attired in the usual Eskimo dress of skins; objects of interest and curiosity, but not "things of beauty."

Hans led his visitors, over rough rocks and through deep drifts of snow, to his rude hut, which stood on the cold hill-top, about two hundred feet above the sea-level. An excellent position for a "look-out," but as inconvenient for a hunter as can well be imagined. Here he had watched and waited for many a dreary month; surveying the sea day after day, in the faint hope of discovering some European vessel. But none came; summer passed into winter, and winter lengthened into summer; and still Hans watched and waited, yearning after his southern home and the friends of his youth.

His tent—for it was rather a tent than a hut—was made of seal-skins, and its capacity was scarcely sufficient to accommodate his little family.

Dr. Hayes asked him if he would accompany the expedition.

"Yes."

Would he take his wife and baby?

"Yes."

Would he go without them?

"Yes."

This last answer reveals the curious unimpressionableness of the Eskimo, who endures with calmness, nay, even with indifference, those partings which try the heartstrings of the European. It is, perhaps, a result of the constant warfare he maintains against an uncongenial and austere Nature that he comes to regard himself as his first and chief, as almost his only concern. So long as his wife and children surround him, he shows no evident want of affection; but he has no objection to part from them, if the separation will prove to his individual interest.

As Dr. Hayes had no leisure to examine critically into the state of his mind, and as he cherished a conviction that the permanent separation of husband and wife was to be regarded as a painful event, he determined on giving the Eskimo mother the benefit of this conventional suspicion. Both husband and wife, therefore, were carried on board the schooner, as well as their baby, their tent, and all their household goods. The bright-eyed boy and the ancient dame cried to accompany them; but Dr. Hayes had no further room, and was compelled to leave them to the care of their tribe, who, about twenty in number, had discovered the schooner, and with a merry shout had come across the hill. After bestowing upon them some useful gifts, Dr. Hayes returned to his vessel.

He adds that Hans was the only unconcerned person in the party. At a later period the thought crossed his commander's mind that he would by no means have been displeased had wife and child been left to the charity of their savage kin: while Dr. Hayes had abundant reason, during the course of the expedition, to wish that he had left the selfish and indolent Eskimo to linger in his seal-skin tent among the hills and rocks of Cape York.

The same traveller describes the hunting equipment of a party of Eskimos setting out in pursuit of bears.

First, the dogs. These were picketed, each team separately, on a convenient area of level ground; and on the approach of Dr. Hayes and his companions they sprang up from the knotted heap, in which they had been lying through the night, with a wild, fierce yell, which died away into a low whine and impatient snarl. They evidently were hungry, and their masters seemed desirous of feeding them; for, going to their sledges, each one brought up a flat piece of something which looked singularly like plate-iron, but, upon examination, was found to be walrus-hide, three-quarters of an inch thick, and frozen intensely hard. Throwing it upon the snow a few feet in advance of their respective teams, they drew their knives from their capacious boots, and attempted to cut up the skin; but its hardness defied all their efforts, and Dr. Hayes had to fetch hatchet and saw before the work of division could be completed.

During the few minutes thus occupied, the dogs had become almost frantic. They endeavoured to break loose; pulling on their traces, running back and springing forward, straining and choking themselves until their eyes shot fire, and the foam flew from their mouths. The sight of food had stimulated their wolfish passions, and they seemed ready to eat each other. Not a moment passed that two or more of them were not flying at each other's throats, and, grappling together, rolled, and tossed, and tumbled over the snow.

The Eskimos looked on apparently unconcerned, except when there appeared a risk of one of the dogs being injured, and then they secured a temporary calm by uttering an angry nasal "Ay! Ay!"

When at length the food was thrown, the dogs uttered a greedy scream, which was followed by a moment's silence while the pieces were falling, then by a scuffle, and the hard, frozen chunks had vanished. How they were swallowed, or how they were digested, was, to the spectator, inexplicable! Enough to say that "the jaws of darkness did devour them up," and calm instantaneously succeeded to the storm.

The Eskimo dog is of medium size, and squarely built; in fact, he is a reclaimed wolf, and exhibits that variety of colour which, after a few generations, generally characterizes tame animals. Gray, which is often seen, was probably at one time the predominating colour. Some of the dogs are black, with white breasts; some are wholly white; others are reddish or yellowish; but, indeed, almost every shade may be seen amongst them. Their skin is covered with a coarse, compact fur, and is much valued by the natives for the purposes of clothing. In the form of the animals the variety is considerable; but the general characters would seem to be a pointed nose, short ears, a cowardly, treacherous eye, and a hanging tail. But exceptions occasionally occur, and one figures in Dr. Hayes' narrative under the name of *Toodlamik*, or, more briefly, *Toodla*.

He differed from his kind in having a more compact head, a less pointed nose, an eye denoting affection and reliance, and an erect, bold, fearless carriage. Dr. Hayes, however, expresses some doubt as to his purity of blood. From the beginning to the end of the cruise he was master of all the dogs that were brought to the ship. In this connection it is worthy of remark, that in every pack one dog invariably attains the mastership of the whole—a kind of major-generalship;

and in each team, one who is master of his comrades, a general of brigade. Once master, always master; but the post of honour is gained at the cost of many a lame leg and ghastly wound, and is held only by doing daily battle against all comers. These could easily gain the ascendancy in every case, but for their own petty jealousies, which often prevent their union for such a purpose. If a combination, however, does happen to be brought about, and the leader is hopelessly beaten, he is never worth anything afterward; his spirit is completely prostrated, the poor fellow pines away, and dies at last of a broken heart.

ESKIMO DOGS.

Toodla, says Dr. Hayes, was a character in his way. He was a tyrant of no mean pretension. Apparently he thought it his special duty to attack every dog, great or small, that was added to the pack: if the animal was a large one, in order, probably, that he might at once be forced to feel that he had a master; if a small one, in order that the others might hold him in the greater awe. It was sometimes quite amusing to see him set off in pursuit of a strange dog, his head erect, his tail curled gracefully over his back; slowly and deliberately he went straight at his mark, with the confident, defiant air of one who recognizes the power and importance of his office.

Leagues and conspiracies were not unfrequently formed against him, induced, no doubt, by a feeling of despair; but he always succeeded in overthrowing them,—not, it is true, without occasional assistance from "without;" for the sailors, who petted him greatly, would sometimes take his part when the struggle was manifestly unequal.

But we must leave the dogs, and turn to the sledge.

This was, in very truth, an ingenious specimen of native mechanical skill. It was made wholly of bone and leather. The runners, which were square behind and rounded upward in front, and about five feet long, seven inches high, and three-fourths of an inch thick, were slabs of bone; not solid, but made up of a number of pieces of various shapes and sizes, dexterously fitted and tightly lashed together. Some of these were not larger than one's two fingers; some were three or four inches square; others were as large as one's hand, and triangular in shape; others, again, were several inches in length, and two or three in breadth. They all fitted into their several places as exactly as the blocks of a Chinese puzzle. Near their margins ran rows of little holes, and through these strings of seal-skin were inserted, by which the blocks were fastened together, until the whole was as firm as a board.

The marvel of the thing is that all these pieces are flattened and cut into the required shape, not with nicely contrived instruments and tools, but with stones. The labour must be immense. The grinding needed to make a single runner must be the work of months. The construction of an entirely new sledge would probably occupy the lifetime of a generation; and hence a vehicle of this kind becomes a family heirloom, and is handed down from father to son, and son to grandson, and is constantly undergoing repair and restoration; a new piece here, another there, until as little remains of the original structure as of the sailor's old knife, when it had had a new blade and a new handle! The origin of some of the Eskimo sledges is lost in the mists of a remote antiquity.

The runners are usually shod with ivory from the tusk of the walrus. The said ivory had likewise been ground flat, and its corners made square, with stones; and it was fastened to the runner by a string looped through two counter-sunk holes. The pieces of which it was composed were numerous; but the surface was wonderfully uniform, and as smooth as glass.

The runners stood about fourteen inches apart, and were fastened together by bones, tightly lashed to them; the bones used being the femur of the bear, the antlers of the reindeer, and the ribs of the narwhal. Two walrus-ribs, lashed one to the after-end of each runner, served as upstanders, and were braced by a piece of reindeer antler, secured across the top.

Having thus disposed of the team and the sledge, we now come to the equipment.

First, one of the Eskimo hunters spread a piece of seal-skin over the sledge, fastening it securely by little strings attached to its margin. On this he placed a small piece of walrus-skin, as a provision for the dogs; a piece of blubber for fuel; and of meat for his own lunch. During his absence he would cook no food, but he would want water; and therefore he carried his *kotluk*, or lamp—namely, a small stone dish; a lump of *mannek* or dried moss, designed for the wick; and some willow-blossoms (*na-owinals*) for tinder. To ignite the tinder, he had a piece of iron-stone and a small sharp fragment of flint.

We may follow him on his route, and ascertain the use he makes of these appliances.

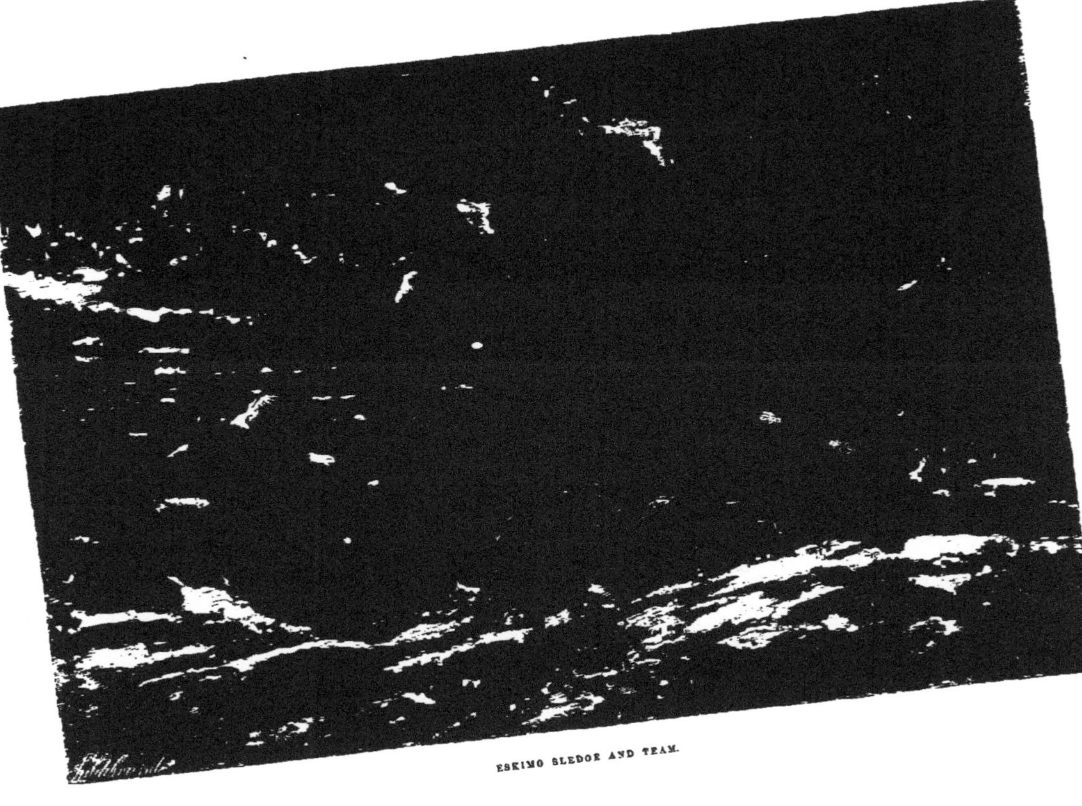

ESKIMO SLEDGE AND TEAM.

When he grows thirsty, he halts; scrapes away the snow until he lays bare the solid ice beneath; and painfully scoops in it a small cavity. Next, he fetches a block of fresh-water ice from a neighbouring berg, lights his lamp, and, using the blubber for fuel, proceeds to place the block on the edge of the cavity. As it slowly thaws, the water trickles down into the hole; and when the Eskimo thinks the quantity collected is sufficient to quench his thirst, he removes the rude apparatus, and, stooping down, drinks the soot-stained fluid. If he feels hungry, he breaks off a few chips from his lump of frozen walrus-beef, cuts a few slices from the blubber, and enjoys his unsatisfactory meal. The inhabitant of the Arctic desert knows nothing of epicurean tastes; and if he did, he has no means of gratifying them.

To return to the equipment. The hunter carried with him an extra pair of boots, another of dog-skin stockings, and another of mittens, to be used in case he should be unfortunate enough to get on thin ice, and the ice should break through.

The entire equipment being placed upon the sledge, he threw over them a piece of bear-skin, which was doubled, so that, when opened, it would be large enough to wrap about his body and protect it from the snow, if he wished to lie down and rest. Then he drew forth a long line, fastened an end of it through a hole in the fore part of one of the runners, ran it across diagonally to the opposite runner, passed it through a hole there, and so continued, to and fro, from side to side, until he reached the other end of the sledge. There he made fast the line; and thus the cargo was secured against all risk of loss from an upset. Next he hung to one upstander a coil of heavy line, and to the other a lighter coil, tying them fast with a small string. The former was his harpoon-line for catching walrus; the latter, for catching seal. His harpoon staff was made from the tusk of the narwhal; measured five feet in length, and two inches in diameter at one end, tapering to a point at the other.

All being ready, the team, consisting of seven dogs, was brought up. The harness was of a very primitive description. It consisted of two doubled strips of bear-skin, one of which was placed on either side of the animal's body, the two being fastened together on the top of the neck and at the breast, so as to form a collar. Thence they passed inside of the dog's fore legs and up along his flanks to the tail, where the four ends meeting together were attached to a trace eighteen feet in length.

The trace was connected with the sledge by a line four feet long, of which one end was attached to each runner. And to the middle of the line a stout string was fastened, running through bone rings at the ends of the traces, and secured by a slip-knot, easily untied—an arrangement designed with the view of ensuring safety in bear-hunting. The bear is hotly pursued until the sledge arrives within about fifty yards; the hunter then leans forward and slips the knot; the dogs, set loose from the sledge, quickly bring the brute to bay. If the knot gets fouled, serious accidents are not unlikely to occur. The hunter vainly endeavours to extricate it, and before he can draw his knife to cut it—supposing he is fortunate enough to have such an instrument—man, and dogs, and sledge are all among the bear's legs, in a huddled and tangled heap, and at the mercy of the enraged monster.

The dogs were cold, and eager to start. In a moment they were yoked to the sledge; the hunter with his right hand threw out the coils of his long whip-lash, with his left he seized an upstander, and propelling the sledge a few paces, he uttered at the same moment the shrill starting-cry, "Ka! ka!—ka! ka!" which sent the dogs in a bound to their places, and away they dashed over the rugged ice. The hunter skilfully guided his sledge among the hummocks, moderating the impetuosity of his team with the nasal "Ay! ay!" which they perfectly understand. On reaching the smooth ice, he dropped upon the sledge, allowed his whip-lash to trail after him on the snow, shouted "Ka! ka!—ka! ka!" to his savage team, and disappeared in as wild a gallop as ever was taken by the demon huntsman of German legend!

It does not appear that the Eskimos have magistrates or laws, yet the utmost good order prevails in their communities, and quarrels are rare. When these do occur, one or other of the dissatisfied parties collects his little store, and migrates to a different settlement. The constitution of their society is rightly described as patriarchal, but the ruler does not seem to be elected: he attains his post by proving his possession of superior strength, address, and courage. As soon as his physical powers give way, or old age enfeebles his mind, he deposes himself, takes his seat in the *oomiak*, or woman's boat, and is relegated by common consent to female companionship. Like all savage tribes, the Eskimos have their mystery-men, or *angekoks*, who resort to the usual deceptions to acquire and retain supremacy, swallowing knives, resorting to ventriloquial artifices, and conversing in a mysterious jargon, unintelligible to "the common herd." They profess to hold intercourse with certain potent spirits, and to employ their agency in rewarding or punishing their dupes; and even the influence of the Christian missionaries has hardly rooted out the belief in the superstitions originated and fostered by these men.

Notwithstanding the hard conditions of their life, and the dreariness of the region which they inhabit, the Eskimos are a cheerful people. They are keenly sensible of the charms of music, though their own vocalization is inconceivably melancholy; and they are partial to many rude pastimes, mostly of a gymnastic character.

Their good nature has been praised by many travellers; but they show the usual inhumanity of the savage towards the aged and infirm. Weakness is no title to the sympathy of the Eskimo; he respects strength, but he utterly disregards and cruelly oppresses the feeble. He is ungrateful towards his benefactors, and in his intercourse with strangers his fidelity can be relied upon only so long as he knows that any breach of faith will be severely punished. He does not steal from his own people, and "Tiglikpok," "he is a thief," is a reproach among the Eskimos as among ourselves; but no shame attaches to him if he robs the white man, though the latter may have loaded him with favours.

If we add that they display a strong affection for their children, and that the children are singularly docile and obedient to their parents, we shall have said enough to assist the reader in forming an accurate conception of the characteristics of the inhabitants of the Eskimo Land.

CHAPTER VIII.

LAPLAND AND THE LAPPS.

LAPLAND, or the Land of the Lapps, which the Lapps themselves call Sameanda or Somellada, forms the north and north-eastern portions of the Scandinavian peninsula, and is divided between Sweden and Russia. Norwegian Lapland includes the provinces of Norrland and Finmark; Swedish, of North and South Bothnia; and Russian, of Kola and Kemi. The last-named has an area of 11,300 square miles, with a population of 9000; Swedish Lapland, an area of 50,600 square miles, with 4000 inhabitants; and Norwegian, an area of 26,500 square miles, with a population of 5000. We are here referring to the number of true Lapps; in each division the population would be largely increased if we included Finns, Russians, Swedes, Norwegians.

Lapland, for nine months in the year, is blighted by the rigour of a winter climate. The summer months, when the sun does not set for several weeks, are July and August; and these are preceded by a brief spring, and followed by even a briefer autumn. Cereals do not thrive higher than the sixty-sixth parallel, with the exception of barley, which is cultivated as far north as the seventieth. The greater part of the country comes within that wooded zone which we described in an earlier chapter, and the forests, consisting of birch, pine, fir, and alder, spread over a very extensive area. On the mosses and lichens which grow abundantly in their shelter, are fed the immense herds of reindeer which constitute the principal wealth of the inhabitants.

The Lapps may almost be regarded as a nation of Lilliputians. Their men seldom exceed five feet in height, while the majority are some inches below that very moderate stature; and the women are even shorter. They are, however, a robust race, with muscular limbs, and unusual girth of body, the circumference of their chest being nearly equal to their height. Their complexion is dark, tawny, or copper-coloured; their dark, piercing, deep-sunken eyes are set very wide apart, so as to communicate a peculiar character to the physiognomy. The wild, strange effect is further increased by the unkempt masses of dark, lank, straight hair which droop on either side of the whiskerless, beardless face. The cheek-bones are prominent, like those of a Celtic Highlander; the nose is flat; the mouth wide, with thin compressed lips. It may be supposed that the Lapps, from these indications, are not models of masculine or feminine beauty; and Dr. Clarke asserts that, when aged, many of them, if exposed in a menagerie, might be mistaken for the long-lost transitional form intermediate between man and ape. And, certainly, there is something repulsive in the constant blinking of eyes rendered sore by the pungent smoke of their huts, or the white glare of the snow, as well as in the expression of obstinacy and low cunning which one reads in every feature.

An aristocrat might be proud of their small and finely-shaped hands; but their arms, like their legs, are disproportionately short, clumsy, and thick. Clumsy, we mean, in shape; certainly not in movement, for the extraordinary flexibility of their limbs is one of the traits by which a Lapp is easily distinguished.

Of the dress of the Lapps it is needless to say much. In winter it consists of bears' skins, in which both male and female wrap themselves up, with the fur outward. In summer the men wear a sort of tunic, the *poesk*, made of coarse light-coloured woollen cloth, depending to the knee, but bound about the waist with a belt or girdle. Their head-gear consists of a kind of fez, made of wool, and adorned with a red worsted band round the rim, and a bright red tassel. Their boots or shoes are cut from the raw skin of the reindeer, with the hair outwards, and they are peaked in shape. They are thin, and they have no lining; but the Lapp defends his feet and ankles from the cold by stuffing the vacant space of the boot with the broad leaves of the *Carex vesicaria*, or Cyperus grass, which he cuts in summer, rubs in his hands, and dries before using. The female costume resembles that of the males, but their girdles are gayer with rings and chains.

The Lapps are a superstitious race. Like all the Norse tribes, they believe in witchcraft; and of old the Lapland witches had a reputation which extended to England, for being able to ward off rain or disperse storms. The English seamen trading to Archangel frequently visited their coast in order to buy a favourable wind.

Many of the Lapps claim the ability to foretell future events, and fall, or pretend to fall, into a trance or ecstasy, during which they see visions, utter prophecies, and unlock the secrets of those who trust to their divination. They also read the fortunes of inquiring dupes by means of a cup of liquor, or by the vulgarest jargon of palmistry. Superstition is the daughter of Ignorance. It is also the sister of Fear, for the superstitious are invariably prone to see supernatural signs and wonders in the appearances of the heavens, or to hear unearthly voices borne upon the midnight wind, and in everything they cannot understand to imagine the presence of some antagonistic power. As the American natives were panic-stricken at the occurrence of an eclipse, so the Lapps are filled with dread when the sky glows with the coruscations of the aurora.

These superstitions prevail in spite of the exertions of priests and schoolmasters. They are nourished in secret even when they are not openly proclaimed; and the Lapp, after listening devoutly to the harangue of his pastor, will return home to offer homage to his saidas, or wooden idols; to cower at the name of Trolls, the evil spirit of the forest; and to be deluded by the artifices of any so-called witch or fortune-teller.

There are Lapps, and Lapps; each, according to the region he inhabits, bearing his distinctive characteristics, and preserving his individual habits. Thus, there are the Fjälllappars, or Mountain Lapps; the Skogslappars, or Wood Lapps; and the Fisherlapps.

From the nature of the country the reader will expect, and will be right in expecting, that the Fjälllappars form the most numerous section. They are the nomads of Lapland, and their mode of life is entirely pastoral. As the Arabs with their flocks move from one oasis to another, or the Tartars with their cattle, so the Lapps migrate from place to place, compelled by the necessity of finding sustenance for their herds of reindeer. The mosses and lichens on which these animals feed are soon exhausted, and some time elapses before the half-frozen soil replaces them. The same cause operates to prevent the Lapps from assembling in large communities. Seldom more than three, four, or five families encamp in the same neighbourhood.

It will not be supposed that the temporary abode of a nomad exhibits any architectural completeness. Their *tuguria*, or huts, are of the rudest construction. They raise a conical framework, composed of the flexible stems of trees, and this they cover with a coarse kind of canvas, and in winter with the skins of reindeer and other animals. No doorway is required, and egress and ingress are provided for by turning up a portion of the canvas at the bottom, so as to form a triangular gap; and the portion so turned up is let down again at night. In the centre of the interior some large stones are piled together for a fireplace, and a square opening in the roof above carries off the smoke, and lets in the light and air—not to say rain, snow, and fog, when these prevail.

The tent or hut we have described generally measures about six feet in diameter, and eighteen to twenty in circumference. It does not exceed ten feet in height. There is no floor, but the ground is covered with reindeer skins, and upon these the inhabitants sit or crouch by day, and huddle themselves up at night. The household utensils, implements, and weapons are suspended from the sides of the hut; and the clothing of the family, no very extensive stock, is preserved in a chest.

On a shelf or platform, raised high above the reach of dogs and wolves, between two neighbouring trees, the Lapp keeps his store of dried reindeer flesh, and cheese, and curds; for his diet is as plain as his general habit of living. His herd of reindeer he puts up at night, or when they are required for milking, in a large enclosure, about four hundred to five hundred feet in circuit, formed by a barrier of posts and stumps of trees, supporting a row of horizontal poles. Against the latter birch poles and branches of trees are placed diagonally, forming a kind of abattis, which is found to be a sufficient security against the attacks of wolves.

It is said that the milking of a herd of reindeer affords a lively and picturesque spectacle. When they have been driven within the area, and all the outlets closed, a Lapp, selecting a long cord or thong, twists both ends round his left hand, and then in his right gathers the thong itself in loose coils. Fixing on a reindeer, he flings the coils over its antlers. Sometimes the latter offers no resistance; but generally, on feeling the touch of the thong, it darts away, and its pursuer, in order to secure it, is called upon for the most vigorous efforts. And the scene is animated indeed, when half-a-dozen reindeer, pursued by as many Lapps, sweep round and round the enclosure, until the former are finally overcome, or, as now and then happens, wrest the cord from the hands of the discomfited Lapp, and leave him prostrate on the ground. When the animal is secured, his master takes a dexterous hitch of the thong round his muzzle and head, and then fastens him to the trunk of a prostrate tree. The operation of milking is performed by both men and women.

As soon as the pasture in the neighbourhood is exhausted, the encampment is broken up, and the little company migrate to some fresh station. The rude tuguria are dismantled in less than half an hour, and packed with all the household furniture on the backs of the reindeer, who, by long training, are inured to serve as beasts of burden. On the journey they are bound together, five and five, with leather thongs, and led by the women over the mountains; while the father of the family precedes the march to select a suitable site for the new encampment, and his sons or servants follow with the remainder of the herd.

As spring verges upon summer, the Lapps abandon their mountain pastures, and move towards the shore. No sooner do the reindeer scent the keen sea-air than, breaking loose from all control, they dash headlong into the briny waves of the fiord, and drink long draughts of the salt sea-water. The Lapps consider this sea-side migration essential to the health of their herds. When summer reaches its meridian, and the snow melts, they return to the pleasant mountain-solitudes, ascending higher and higher, according to the increase of temperature. Then, on the approach of winter, they retire into the woods, where their great difficulty is to defend their herds and themselves from the attacks of the wolves. In this incessant warfare they derive much

REINDEER IN LAPLAND.

assistance from the courage of their dogs. Those are about the size of a Scotch terrier, with long shaggy hair, and a head bearing a curiously close resemblance to that of a lynx.

In the winter the Lapp accomplishes his journeys either by sledging or skating.

Their skates are not exactly things of beauty, but they answer their purpose admirably. One is as long as the person who wears it; the other is about a foot shorter. The feet of the wearer are placed in the middle, and the skates, or *skidas*, fastened to them by thongs or withes. They are made of fir-wood, and covered with the skins of reindeer, which check any backward movement by acting like bristles against the snow. It is astonishing with what speed the Lapp, thus equipped, can traverse the frozen ground. The most dexterous skater on the canals of Holland could not outstrip him. He runs down the swiftest wild beasts; and the exercise so stimulates and warms his frame that, even in midwinter, when pursuing one of these lightning-like courses, he can dispense with his garment of furs. When he wishes to stop, he makes use of a long pole, which is provided with a round ball of wood near the end, to prevent it from sinking too deep into the snow.

He is no less expert as a sledger. His vehicle, or *pulka*, is fashioned like a boat, with a convex bottom, so as to slip over the snow with all the greater ease; the prow is sharp and pointed, but the hind part flat. Perhaps it may better be compared to a punt than a boat. At all events, in this curious vehicle the Lapp is bound and swathed, like an infant in its cradle. To preserve its equilibrium, he trusts to the dexterity with which he moves his body to and fro, and from side to side, as may be needed; and he guides it by means of a stout pole. His steed, a reindeer, is fastened to it by traces attached to its collar, and connected with the fore part of the sledge; the reins are twisted round its horns; and all about its trappings are hung a number of little bells, in the tintinnabulation of which the animal greatly delights. Thus accoutred, it will perform a journey of fifty or sixty miles a day; sometimes travelling fifty miles without pause, and with no other refreshment than an occasional mouthful of snow.

TRAVELLING IN LAPLAND.

With wonderful accuracy the Lapp will guide himself and his steed through a seemingly labyrinthine wilderness, when the usual signs and characters of the landscape are buried deep in snow. But his memory is tenacious, and a blighted tree, or a projecting crag, or a clump of firs, affords him a sufficient indication of the correctness of his course. He frequently continues his rapid journey throughout the night, when the moon invests the gleaming plains with a strange brilliancy, or the aurora fills both earth and heaven with the reflection of its wondrous fires.

A French traveller, M. de Saint-Blaize, is of opinion that the Lapps, like all savage and semi-civilized races, are rapidly diminishing in numbers. Yet this diminution is hardly owing to the conditions under which they live. Their life, to the civilized European, seems severe and almost intolerable; but though it is marked by privation and fatigue, it is not without its charms. It is free and independent, and without anxiety. As for the privation and fatigue, the Lapp is hardly conscious of them, because his capacity of endurance is great, and he is accustomed to them from his earliest years. Temperate, active, and inured to exertion, his physical frame is wonderfully vigorous, and he knows nothing of the majority of maladies which afflict the dweller in cities. One terrible disease, indeed, he does not escape, and this may have had much to do with their decline,—the smallpox. Otherwise, they are a healthy as well as a hardy race. If during a journey a Lapp woman gives birth to a child, she places the new-born in a frame of hollow wood, in which a hole has been cut to receive the little one's head; then slings this rude cradle on her back, and continues her march. When she halts, she suspends the infant and its cradle to a tree, the wirework with which it is covered affording a sufficient protection against wild beasts.

Professor Forbes, however, describes a more comfortable cradle, which is cut out of solid wood, and covered with leather, in flaps so arranged as to lace across the top with leathern thongs; the inside is lined with reindeer moss, and a pillow, also of reindeer moss, is provided for the head of the infant, who fits the space so exactly that it can stir neither hand nor foot.

The Lapp is a bold hunter, and will encounter the bear single-handed. Like the Siberian, he entertains a superstitious reverence for this powerful animal, which he regards as the wisest and most acute of all the beasts of the field, and supposes to know and hear all that is said about it; but as its fur is valuable and its flesh well-favoured, he does not refrain from pursuing it to the death, though careful, so to speak, to kill it with the highest respect.

Early in winter the bear retires to a rocky cave, or a covert of branches, leaves, and moss, and there remains, without food, and in a state of torpidity, until the spring recalls him to active life. After the first snowfall, the Lapp hunters seek the forest, and search for traces of their enemy. These being found, the spot is carefully marked, and after a few weeks they return, arouse the slumbering brute, and stimulate it to an attack; for to shoot it while asleep, or, indeed, to use any weapon but a lance, is considered dishonourable.

Hogguer, whose narrative is quoted by Hartwig, accompanied a couple of Lapps, well armed with axes and stout lances, on one of these dangerous expeditions. When about a hundred paces from the bear's den, the party halted, and one of the Lapps advanced shouting, and his comrades made all the din they could. He ventured within twenty paces of the cavern, and then threw stones into it. For awhile all was quiet, and Hogguer began to think they had come upon an empty den; but suddenly an angry growl was heard.

The hunters now renewed and redoubled their clamour, until slowly, like an honest citizen roused from his virtuous sleep by a company of roisterers, the animal came forth from his lair.

At first he seemed indifferent and lethargic; but, catching sight of his nearest enemy, he was filled with rage, uttered a short but terrible roar, and rushed headlong upon him. The Lapp, with his lance in rest, awaited the onset calmly, while the bear, coming to close quarters, reared himself on his haunches, and struck at his antagonist with his fore paws.

To avoid these powerful strokes, the daring huntsman crouched, and then, with a sudden spring, drove his lance, impelled by a sturdy arm, and guided by a sure eye, into the creature's heart.

The victor escaped with only a slight wound on the hand, but the marks of the bear's teeth were found deeply impressed on the iron spear-head.

According to an old custom, the wives of the hunters assemble in one of their huts, and as soon as they hear them returning, raise a loud discordant chant in honour of the bear. When the men, loaded with their booty of skin and flesh, draw near, it is considered necessary to receive them with words of reproach and insult, and they are not allowed to enter through the door; they are compelled, therefore, to obtain admission through a hole in the wall. But when the animal's *manes* have been thus propitiated, the women are not less eager than the men to make the most of its carcass; and after the skin, fat, and flesh have been removed, they cut up the body, and bury it with great ceremony, the head first, then the neck, next the fore paws, and so on, down to the animal's "last,"—its tail. This is done from a wild belief that the bear rises from the dead, and if it has been properly interred, will kindly allow itself to be killed a second time by the same hunter!

The principal article of food of the Lapps is reindeer venison. This they boil, and it supplies them both with meat and broth. In summer they vary their bill of fare with cheese and reindeer milk; and the rich eat a kind of bread or cake, baked upon hot iron plates or "girdles." For luxuries they resort to brandy and tobacco; and these are not less appreciated by the women than by the men. As for the latter, they are never seen without a pipe, except at meals; and the first salutation which a Lapp addresses to a stranger is a demand for "tabak" or "braendi." Dr. Clarke tells us that on paying a visit to one of their tents, he gave the father of the family about a pint of brandy, and as he saw him place it behind his bed, near the margin of the tent, he concluded it would be economically used. In a few minutes the daughter entered, and asked for a dram, on the ground that she had lost her share while engaged upon domestic duties outside. The old Lapp made no reply, but slily crept round the exterior of the tent until he came to the place where the brandy was concealed. Then, thrusting in his arm, he drew forth the precious bottle, and emptied its contents at a draught.

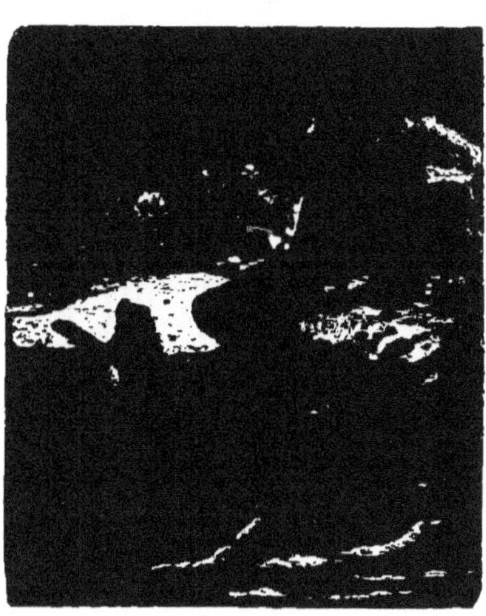

FISHER LAPPS.

We find no great difference of habits existing between the Mountain Lapp and the Skogs or Forest Lapp, except that the latter takes up fishing as a summer pursuit, and devotes the winter months to his herds and the chase. But in course of time his herds demanding more attention than he can give to them, he is transformed into a Fisher Lapp, who dwells always upon the sea-coast, and is at once the filthiest and least civilized of the race. He resembles the Mountain Lapp in his love of tobacco and brandy. He differs from him in never migrating, and in wholly abandoning the pastoral life.

A picture of what the artists call a Lapland "interior," of the domestic economy of a Lapp hut, is painted for us by the author of a recent book of travel, entitled "Try Lapland."

After a long day's journey, in the neighbourhood of Lake Randejaur, weary and cold, he and his companions came upon a small hut, and had visions of obtaining a night's rest; but a closer acquaintance with the hut convinced them that such a proceeding would be undesirable.

For, knocking at the door, and pulling up the latch, they entered, to see before them a family scene!

In an inconceivably dirty room stood a still dirtier beldame, making coffee. Her husband, an old man of seventy, sat on one side; while a hideous, deformed little Lapp, whether man or woman they could hardly tell, squatted on the floor on the other, in full costume, consisting of high-peaked blue cloth cap, and reindeer-skin dress, ornamented with beads and spangles. Her face was brown as a berry, long lanky black hair streamed down her cheeks; and, staring at the intruders, she begged for "penge" (money). Two young men were snoring in one bed, and two boys in another placed opposite to it, each being covered with a few reindeer-skins.

The entrance of the strangers aroused the sleepers to give one hasty look, and then they snored again.

The lady of the house offered coffee; and though everything looked so dirty as to create a positive feeling of disgust, the travellers could not afford to be particular, and accepted her offer, which put her in a perfect ecstasy of delight.

Quickly she scuttled off to the well for water, and, filling her kettle, set to work to roast fresh coffee.

The old man got up and endeavoured to rouse the sleepers, when he understood that the strangers were in immediate want of boats and rowers.

Leaving him to make the necessary preparations, they went out to take a look at the surrounding scenery; and returning in a quarter of an hour, expected to find them preparing the boats, which lay two or three hundred yards off. But, to their surprise, not the slightest change had occurred in the position of the sleepers; and, after drinking their coffee out of the one cup the Lapps possessed, they grew impatient, and stormed at the young men, trying even to pull them out of bed—but they would not budge.

"The father," says our authority, "who protested great love for the English, but turned out the biggest rascal we had come across, was as anxious as we were that his sons should get up and row us;—but not a bit of it! He told us that they had been out three days and three nights on the Fells, and were thoroughly exhausted. What was to be done, we could not think. It was getting serious; we certainly could not sleep in this dreadful hole, and there was no other shelter near.

"Money had no power: though I showed the almighty dollar to the weary slumberers, (they had surely never been in America!) they turned away with a grunt.

"Then, O happy thought, I recollected the *brandy;* and bringing my keg to the bedside, I tapped it, and offered them a glass if they would get up. This was quite another thing; they yawned, stretched their limbs, and stood upon the floor. Poor fellows! we then saw how ill and fagged they looked, though they were splendid specimens of the human race.

"Pouring a glass of the fiery compound down their throats, they put on their coats, and followed us like sleepy dogs; but in a few moments were rowing us like heroes."

All travellers agree in bearing witness to the passion of the Lapps for alcoholic liquors. If we could spare our apostles of temperance and advocates of Good Templarism, which, alas! we cannot afford to do, few better fields could be found for their admirable labours than Lapland.

Captain Hutchinson, however, has more pleasant experiences to relate, and more agreeable "interiors" to sketch, than the preceding. Let us accompany him, for instance, on a visit to the island of Bjorkholm.

The settlement here is very small, consisting of only two or three houses, and a few barns and sheds. The inhabitants, after the usual manner of the Lapps, support themselves by fishing in summer, and by the reindeer in winter. Not a tree or shrub grows upon the island; only grass.

The hostess, on this occasion, was an active, good-natured little woman, not more than four feet high, who flew to and fro with a really wonderful agility. At one moment she was mounted on the dresser, searching for forks and spoons; at another, almost buried in a deep box, diving for sheets and table-cloth. Crockery was decidedly scarce; and a china slop-basin, with a wreath of prettily painted little flowers round the margin, had really a hard time of it.

It was first presented to Captain Hutchinson and his party for the purpose of washing their hands; at supper it appeared filled with chocolate; in the morning it reappeared as their joint washing-basin.

However, the little Lapp entertained them right royally, with hot kippered salmon, pancakes, dried reindeer, and eggs.

The beds were very comfortable, the mattresses of hay, with the whitest of sheets. And though the hostess and her family seemed very poor, relics of former grandeur were visible in the silver spoons, teapot, goblet, and cream-jug.

A recent writer observes that the inferiority of the Lapp race is as conspicuous from the intellectual as from the physical point of view. This is evident from the most cursory glance at their lives and manners. The Lapp is, on the whole, a simple, timid, regular, honest creature. To his great defect we have already adverted,—that excessive partiality for strong liquors, which would be sufficient to bring about the annihilation of his race within a more or less limited period, even if his days were not numbered from every other concurrent cause. He is essentially nomadic. He is perfectly free and independent throughout the solitary wastes which extend from the North Cape to the sixty-fourth degree of latitude; he plants his tent where he pleases, generally close to a wood or lake; and he moves on when the moss all around it has been eaten up. Such a mode of life is, of course, incompatible with the progress of Swedish, Norwegian, and even Finlandish civilization, which, year by year, curtails the territory given up to the migration of the nomadic Lapps.

There is about the life of the Lapps, in summer, says Count D'Almeida, a certain charm of independence, which might prove seductive to certain minds, weary of civilization and unwitting of mosquitoes. But in winter, no being of any other race could with impunity endure such privations and sufferings as they undergo. They are compelled to keep a careful watch upon their herds, which are in constant danger from the snow-storms and the wolves. In the hard frosts, when the snow is upwards of three feet in depth, they are compelled to dig it up with their axes, so as to obtain access for their reindeer to the moss, which constitutes their only food in winter. Their vigorous constitutions and their power of enduring privation and climatic rigour, explain how it was that man, in the Glacial Age, though without any of the appliances of civilization, could endure its tremendous severity. What the Lapps can bear in point of toil and want is almost incredible. They suffer, and are strong, in a sense the poet never contemplated. It frequently happens that they are surprised by a snow-hurricane; they sleep on the ground, covered with snow-flakes, which, on awaking, they simply shake off, and pursue their way. In an excess of cold which would chill our blood, even if we were running at the top of our speed, they will fall, in a fit of intoxication, on the ground, and lie there with impunity for hours. It is said that in mid-winter, women, suddenly seized with the pains of childbirth while on the road, are delivered in the snow, without any ill result, either to them or their offspring.

But, as the same writer remarks, human strength cannot exceed certain limits. The Lapp ages early in life, and dies young. When he attains an advanced age, his fate is still more lamentable. It is said that if an old man falls sick while a tribe is accomplishing one of its customary migrations, his children frequently abandon him,—leaving him with some provisions at the foot of a tree, or on the bank of a stream, with the terrible prospect before him of dying of starvation, or falling a prey to wild beasts. The Lapp is always poor even when he may be called rich; for it is calculated that to maintain a family of four persons, a herd of fully four hundred reindeer is necessary, representing a capital of about £160.

The Lapp dialect is described as resembling the Finnish. When we remember that the Lapps and the Quénes, or Finns, wear a similar costume, are distinguished by very similar customs, and that the two people call themselves by the same generic name, Suomi, we can understand why some travellers persist in regarding them as sprung from the same common stock. But a careful investigation shows the absolute distinctness of the Lapps from the Finns, notwithstanding this similarity of name and language—a similarity due, as in many other countries, to the influences of conquest or colonization. Some ethnologists, and among them M. D'Omalius, include the Finns among the white, or Caucasian race, and leave the Lapps among the inferior branches of the great Mongol family. It seems certain that a greater difference exists between the Quénes and the Lapps of Northern Norway than between the Quénes and the Scandinavians of the same region.

The Quénes have adapted themselves completely to sedentary and agricultural habits, while the Lapps, as yet, have not made a single advance in the direction of raising themselves above a pastoral and nomadic life. On the other hand, Finns constantly intermarry with the Swedes or Norwegians; while unions between Lapps and Scandinavians, or even between Lapps and Finns, are regarded throughout the entire country as monstrous anomalies. Lastly: laying aside the arguments founded upon the physical conformation of the Lapps and the Finns, an important historical consideration seems to prove their distinct co-existence from a period far anterior to the settlement of the Suiones and the Goths in the peninsula; it is that in the Finnish mythology we constantly meet with legends of battles between dwarfs and giants. It is impossible that these can refer to the warfare between the Finns and the Scandinavians, for the latter were of the same stature as the former; and it is in comparison with the Lapps only that the Finns could relatively be called *giants*.

We borrow from Count D'Alviella a few particulars relating to the stationary Lapps, who inhabit the region of West Bothnia, or Westerbotten, a long, narrow strip of land dividing the Gulf of Bothnia from Lapland proper. These Lapps seem to be the product of a mixture of races in which the Scandinavian element predominates. They are of an ordinary stature, robust, with regular features, light hair, and clear gray eyes.

The country in which they dwell has a strange, an original, but a monotonous character.

It is its monotony which wearies the traveller, though at first he will be impressed by its fresh yet severe beauty. The forests of birch and fir seem endless, and the great lakes in their depths fatigue the eye with their wastes of cold, drear water. Occasionally, however, the traveller comes upon a smiling plain, enamelled with myosotis, and brightened by a silver-shining, music-murmuring stream. Here and there the wood is thinner, and lean cows may be seen feeding among the half-stripped stems. Next comes a clearing, where the forest has been swept away by fire; a clearing with fields of rye and barley; a palisade enclosure, and a group of châlets, with a comparatively spacious and undilapidated building in the centre.

These *gârds*, as they are called, closely resemble each other throughout the North. Neither material nor space is begrudged to the West Bothnian architects. Even the smallest farm comprises three or four buildings, which generally form a square on the four sides of an inner court. These buildings—how unlike the wretched, filthy hut of the nomadic Lapp!—comprise three living-rooms, kitchen, and stables; and are divided from each other only by a partition of horizontally-laid planks, the interstices being filled up by moss. The furniture is simple, convenient, suitable, and shining with cleanliness, like a Dutch kitchen. Around the hearth is hung a series of brightly-coloured prints, representing either a Scriptural scene or events in the life of an illustrious personage,—King Charles XV., or the bishop of the diocese, side by side with the universal legendary figures, Napoleon I. and Garibaldi. Close by stands the old hereditary locker, in which the husband accumulates his money and the wife deposits her trinkets; to the wall is suspended a complete trophy of knives, pipes, belts with silver buckles, sledge-bells, and a whip with a carved horn handle. The whole scene is one of order and the proprieties of family life.

All these dwellings, it may be added, do not wear the same aspect of prosperous neatness; but even where poverty is present, it is unaccompanied by that sullen gloom and melancholy squalidness which, in other countries, is the painful indication and result of long-endured privation. And here, we must also remember, poverty and famine are not always inseparable companions. The shadow of hunger frequently darkens the rich man's door, and a man might perish for want of food on a sack of gold. One winter, the wealthiest members of the community were reduced to the necessity of eating bread made of bark mixed with moss.

Still, we see how wide a difference separates the stationary from the nomadic Lapp, and how impossible it is for a wandering population to acquire or appreciate the comforts of civilized life. A pastoral race, in the present age of the world, is, and must be, a decaying, because a barbarous race. If it touches the borders of civilization, it is only to become infected with its vices, and thus to hasten its inevitable decay.

CHAPTER IX.

THE SAMOJEDES AND OTHER TRIBES OF ARCTIC ASIA.

THE Samojedes are the immediate neighbours of the Lapps. Like them, they are nomades; but they are even less civilized, and have profited less by the arduous and enthusiastic labours of the Christian missionaries. They range over the forests and stony tundras of Northern Russia and Western Siberia; driving their reindeer herds from the banks of the Chatanga to the icy shores of the White Sea, or hunting the wild beasts in the dense woods which extend between the Obi and the Yenisei.

They are sunk far deeper than the Lapps in a coarse and debasing superstition. It is true that they believe in a supreme deity—Num, or Jilibeambaertje, who resides in the air, and, like the Greek Zeus, sends down thunder and lightning, rain and snow; and they evince that latent capacity for poetical feeling which is indicated even by the most barbarous tribes in their description of the rainbow as "the hem of his garment." They regard him, however, as so elevated above the world of man, and so coldly indifferent to humanity, that it is useless to seek to propitiate him either by prayer or sacrifice; and they have recourse, accordingly, to the inferior gods,—who, as they believe, have the direction of human affairs, and are influenced by incantations, vows, or special homage.

The chief of all the Samojede idols is still supposed to consecrate with its presence, as in the days of the adventurous Barentz, the bleak and ice-bound island of Waigatz. It is a block of stone, pointed at the summit, and bearing some rude resemblance to a human head, having been fashioned after this likeness by a freak of nature. This has formed the model for the Samojede sculptors, who have multiplied its effigy in wood and stone; and the idols thus easily created they call *sjadæi*, because they wear a human (or semi-human) countenance (*sja*). They attire them in reindeer-skins, and embellish them with innumerable coloured rags. In addition to the *sjadæi*, they adopt as idols any curiously contorted tree or irregularly shaped stone; and the household idol (Hahe) they carry about with them, carefully wrapped up, in a sledge reserved for the purpose, the hahengan. One of the said penates is supposed to be the guardian of wedded happiness, another of the fishery, a third of the health of his worshippers, a fourth of their herds of reindeer. When his services are required, the Hahe is removed from his resting-place, and erected in the tent or on the pasture-ground, in the wood or on the river's bank. Then his mouth is smeared with oil or blood, and before him is set a dish of flesh or fish, in return for which repast it is expected that he will use his power on behalf of his entertainers. His aid being no longer needed, he is returned to the hahengan.

Besides these obliging deities, the Samojede believes in the existence of an order of invisible spirits which he calls *Tadebtsios*. These are ever and everywhere around him, and bent rather upon his injury than his welfare. It becomes important, therefore, to propitiate them; but this can be done only through the interposition of a *Tadibe*, or sorcerer; who, on occasion, stimulates himself into a state of wild excitement, like the frenzy of the Pythian or Delphic priestess. When his aid is invoked by the credulous Samojede, his first care is to attire himself in full magician's costume—a kind of shirt, made of reindeer leather, and hemmed with red cloth. Its seams are trimmed in like manner; and the shoulders are also decorated with red cloth tags, or epaulettes. A piece of red cloth is worn over the face as a mask, and a plate of polished metal gleams upon his breast.

SAMOJEDE HUTS ON WAIGATZ ISLAND.

Thus costumed, the Tadibe takes his drum of reindeer-skin, ornamented with brass rings, and, attended by a neophyte, walks round and round with great stateliness, while invoking the presence of the spirits by a discordant rattle. This gradually increases in violence, and is accompanied by the droning intonation of the words of enchantment. The spirits in due time appear, and the Tadibe proceeds to consult them; beating his drum more gently, and occasionally pausing in his doleful chant,—which, however, the novice is careful not to interrupt,—to listen, as is supposed, to the answers of the aerial divinities. At length the conversation ceases; the chant breaks into a fierce howl; the drum rattles more and more loudly; the Tadibe seems under a supernatural influence; his body quivers, and foam gathers on his lips. Then suddenly the frenzy ceases, and the Tadibe utters the will of the Tadebtsios, and gives advice how a straying reindeer may be recovered, or the disease of the Samojede worshipper relieved, or the fisherman's labour rewarded with an abundant "harvest of the sea."

The office of the Tadibe is usually transmitted from father to son; but occasionally some individual, predisposed by nature to fits of excitement, and endowed with a vivid imagination, is initiated into its mysteries. His morbid fancy is worked upon by long solitary self-communings and protracted fasts and vigils, and his frame by the use of pernicious narcotics and stimulants, until he persuades himself that he has been visited by the spirits. He is then received as a Tadibe with many ceremonies, which take place at midnight, and he is invested with the magic drum. It will be seen, therefore, that the Tadibe, if he deceives others, partly deceives himself. But he does not disdain to have recourse to the commonest tricks of the conjuror, with the view of imposing upon his ignorant countrymen. Among these is the famous rope-trick, introduced into England by the Davenport Brothers, and since repeated by so many professional necromancers. With his hands and feet fastened, he sits down on a carpet of reindeer-skin, and, the lights being put out, invokes the spirits to come to his assistance. Soon their presence is made known by strange noises; squirrels seem to rustle, snakes to hiss, and bears to growl. At length the disturbance ceases, the lights are re-kindled, and the Tadibe steps forward unbound; the spectators, of course, believing that he has been assisted by the Tadebtsios.

As barbarous, says Dr. Hartwig—to whose pages we are here indebted—as barbarous as the poor wretches who submit to his guidance, the Tadibe is incapable of improving their moral condition, and has no wish to do so. Under various names,—*Schamans* among the Tungusi, *Angekoks* among the Eskimos, *Medicine-men* among the Crees and Chopewyans,—we find similar magicians or impostors assuming a spiritual dictatorship over all the Arctic nations of the Old and the New World, wherever their authority has not been broken by Christianity or Buddhism; and this dreary faith still extends its influence over at least half a million of souls, from the White Sea to the extremity of Asia, and from the Pacific to Hudson Bay

The Samojedes, like the Siberian tribes, offer up sacrifices to the dead, and perform various ceremonies in honour of their memory. Like the North American Indians, they believe that the desires and pursuits of the departed continue to be the same as they were on earth; and hence, that they may not be in want of weapons or implements, they deposit in or about their graves a sledge, a spear, a cooking-pot, a knife, an axe. At the funeral, and for several years afterwards, the kinsmen sacrifice reindeer over the grave. When a prince dies, a Starschina, the owner, perhaps, of several herds of reindeer, his nearest relatives fashion an image, which is kept in the tent of the deceased, and to which as much respect is paid as was paid to the man himself in his lifetime. It occupies his usual seat at every meal; every evening it is undressed, and laid down in his bed. For three years these honours are kept up, and then the image is buried, from a belief that the body by that time must have decayed, and lost all recollection of the past. Only the souls of the Tadibes, and of those who have died a violent death, are privileged with immortality, and hover about the air as disembodied spirits.

A SAMOJEDE FAMILY.

The Samojedes are scattered—to the number of about a thousand families—over their wild and inhospitable region. Ethnologists generally consider them to have a common origin with the Finns of Europe. In stature they are somewhat taller than the Lapps, and their colour is more of a tawny. The marked features of their countenance recall the Hindu type. The forehead is high, the hair black, the nose long, the mouth well-formed; but the sunken eye, veiled by a heavy lid, expresses a cruel and perfidious nature. The manners of the Samojedes are brutal; and in character they are fierce and cunning. They are shepherds, hunters, traders—and when opportunity serves, robbers. Like the other Arctic peoples, they clothe themselves in reindeer-skins. They shave off their hair, except a tolerably large tuft which they allow to flourish on the top of the head, and they pluck out the beard as fast as it grows. The women decorate their persons with a belt of gilded copper, and with a profusion of glass beads and metallic ornaments.

Continuing our progress eastward, we come to the Ostiaks, a people spreading over the northernmost parts of Siberia, from the Oural Mountains to Kamtschatka.

Some interesting particulars of their habits and customs are recorded by Madame Felinska, a Polish lady whom the Russian Government condemned to a long exile in Siberia.

One day, when she was seeking a pathway through a wood, she fell in with a couple of Ostiaks on the point of performing their devotions. These are of the simplest kind: the worshipper places himself before a tree (the larch, by preference) in the densest recess of the forest, and indulges in a succession of extravagant gestures and contortions. As this form of worship is prohibited by the Russian Government, the Ostiak can resort to it only in secret. He professes, indeed, to have accepted Christianity, but there is too much reason to fear that the majority of the race are still attached to their heathen creed.

Nearly every Ostiak carries about his person a rude image of one of the deities which he adores under the name of *Schaïtan*; but this does not prevent him from wearing a small crucifix of copper on his breast. The *Schaïtan* is a rough imitation of the human figure, carved out of wood. It is of different sizes, according to the various uses for which it is intended: if for carrying on the person, it is a miniature doll; but for decorating the Ostiak's hut an image can be had on a larger scale. It is always attired in seven pearl-embroidered chemises, and suspended to the neck by a string of silver coins. The wooden deity occupies the place of honour in every hut,—sometimes in company with an image of the Virgin Mary or some saint,—and before beginning a repast the Ostiaks are careful to offer it the daintiest morsels, smearing its lips with fish or raw game; this sacred duty performed, they finish their meal in contentment.

The priests of the Ostiaks are called *Schamans*; their immense influence they employ to promote their own personal interests, and maintain the meanest superstitions.

In summer the Ostiak fixes his residence on the banks of the Obi or one of its tributaries. It is generally square in form, with low stone walls, and a high pointed roof made of willow-branches, and covered with pieces of bark. These having been softened by boiling, are sewn together so as to form large mats or carpets, which are easily rolled up and carried from place to place. The hearth is in the centre; it consists of a few stones set round a cavity in the soil. Here the Ostiak lives; supporting himself on fish, which he frequently eats without cooking—and purchasing a few occasional luxuries, such as tobacco and drink, with the salmon and sturgeon caught by his dexterity.

In winter he withdraws into the woods, to hunt the sable or the squirrel, or to pasture the herds of reindeer which some of them possess. He builds his jurt on a small eminence near the bank of a stream, but out of reach of its spring inundations. It is low, small, squalid; its walls plastered with clay; its window made of a thin sheet of ice.

The Ostiaks are generally of small stature, dark-complexioned, and with black hair, like the Samojedes; but this is not invariably the case. They seem to belong to the same family as the Samojedes and Finns. They are honest, good-natured, inert, and extremely careless and dirty in their habits; though it may be conceded that their huts are not filthier than the "interiors" of the Icelandic fishermen. Their women are not much better treated than African slaves, and are given in marriage to the highest bidder. The price necessarily varies according to the condition of the parent; the daughter of a rich man sells for fifty reindeer, of a poor man for half-a-dozen dried sturgeon and a handful of squirrel-skins.

JAKUT HUNTER AND BEAR.

The Ostiaks and the Samojedes are great hunters of the white bear. It is the same with the Jakuts (or Yakouts), a people dwelling near the Bouriats, and, like them, approximating to the Mongol type. Their object in the chase, however, is not always to kill the animal, but to take it alive. Madame Felinska asserts that, one day, she saw a considerable herd of bears conducted to Bérézov, like a herd of tame cattle, and apparently quite as inoffensive. She does not inform us, however, by what means they had been reduced to such a desirable state of subjection. Frequently the Ostiaks and the Jakuts attack the white bears body to body, without any other weapon than a hatchet or long cutlass. They require to strike their formidable antagonist with immense vigour, and to slay it at the first blow, or their own danger is extreme. Should the hunter miss his stroke, his sole resource is to fling himself on the ground and lie motionless, until the bear, while smelling his body and turning him over, incautiously offers himself again to his attack.

We now reach the peninsula of Kamtschatka. In area it is equal to Great Britain, and its natural resources are abundant; yet, owing to the ravages of small-pox, and excessive brandy-drinking, its population does not exceed seven or eight thousand souls. Its climate is much milder than that of the interior of Siberia, being favourably affected by the warm breezes from the sea; and though cereals do not flourish, its pasture-grounds are rich and ample, and its herbaceous vegetation is exceedingly abundant.

The fisheries of Kamtschatka enjoy a well-deserved reputation. In spring the salmon ascend its rivers in such astonishingly numerous legions, that if you plunge a dart into the water you will surely strike a fish; and Steller asserts that the bears and dogs in this fortunate region catch on the banks with their paws and mouths more fish than in less favoured countries the most skilful anglers can ensnare by all the devices of piscatorial science. Hermann also refers to the teeming myriads of the Kamtschatka waters. In a stream only six inches deep he saw countless hosts of chackos (*Slagocephalus*), two or three feet in length, partly stranded on the grassy banks, partly attempting to force a passage through the shallows.

The coasts of Kamtschatka swarm in like manner with aquatic birds, which roost and breed on every crag and ledge, in every niche and hollow, and at the slightest alarm rise from their resting-places with a whirr of wings and a clamour of voices repeated by a thousand echoes.

The Kamtschatkans display in the pursuit of these birds and their eggs a skill and a daring not inferior to the intrepidity and dexterity of the inhabitants of the Faroe Isles or the Hebrides. Barefooted, and without even the aid of ropes, they venture to descend the most awful declivities, which the foaming waters render inaccessible from below. On the left arm hangs a basket, to be filled with eggs as they advance; in the right hand they carry a short iron hook, with which to drag the birds from their rocky roosts. When a bird is caught, the fowler wrings its neck, slings it to his girdle, and lowers himself still further down the rugged precipice.

The Kamtschatkans are of small stature, but strong-limbed and broad-shouldered. Their cheek-bones are high, their jaws massive, broad, and prominent, their eyes small and black, their noses small, their lips very full. The prevailing colour of the men is a dark brown, sometimes approaching to tawny; the complexion of the women is fairer; and to preserve it from the sun, they embellish it with bears' guts, adhering to the face by means of fish-lime. They also paint their cheeks a brilliant red with a sea-weed.

KAMTSCHATKANS.

Kamtschatka boasts of a very valuable domestic animal in its dog. Mr. Hill is of opinion that he must be considered indigenous to the country, where he roves wild upon the hills, and obtains his existence in exactly the same manner as the wolf. In his nature, both physically and in respect to his temper and disposition, he seems about equally to resemble that tameless animal and the mastiff; yet not altogether in the same manner that might be supposed to arise from the cross breed of the two species, but rather as possessing some of the qualities of both, neither confounded nor modified, but distinctly marked, and perhaps in equal perfection to the same qualities possessed severally by those animals. He is about the size of the ordinary mastiff, and his colour is usually buff or silver-gray, with the several darker or lighter shades of these colours as an invariable basis. In the form of his body, too, he resembles the mastiff, but his head is more like that of the wolf. Still more do we recognize the wolfish character in the eye, which is cruel and furtive, as well as in his habits and disposition. Like his fellow-rover, he sleeps more by day than by night, and he sees better through the scanty light afforded by the stars or moon than in the full radiance of the sun: this has given rise to the same vulgar error concerning his vision which, in Britain, prevails respecting that of the cat,—that he can see in the dark.

If there be any exception, says Mr. Hill, to the distinct manner in which the dog of Kamtschatka possesses the character and qualities of both the wolf and mastiff, it is in regard to his voice, which is heard in loud cries and undistinguishable sounds, something between the bark of the one and the howl of the other.

In all things connected with the labour in which he is engaged, the Kamtschatka dog displays a more than ordinary intelligence. He is very eager to work, and obedient, like the canine species generally, to one master only; but he gives no indications of that attachment which, more or less, in all other species of the dog, enables man to sympathize with them, and sometimes even excites a degree of friendship which not every one of his own species is able to inspire. Thus, every pack or team of dogs must always be driven by the same hand and guided by the same voice, which the whip, and not caresses, has taught them to remember and obey.

With these qualities, the dog becomes in this country a very serviceable animal. Whatever, indeed, our horses and bullocks perform for us here in Britain, if we except carrying us on their backs and ploughing our arable land, the dogs perform for the Kamtschatkans. There is not much employment for them, however, in the summer; and at that season they are allowed to range about and secure their food, which they usually find in the rivers, in the best way they can. Some pains are at all times necessary to keep them in good temper and at peace with their neighbours, whether canine or human. And therefore all Kamtschatkans who keep a team near their houses are careful, when the snow is on the ground, to drive a number of stakes into the earth, or poles set up in the same manner as the frame of a hut or wigwam; and to these the dogs are attached singly or in pairs. But when paired, whether at the stakes or in harness, it is requisite that those yoked together should be not only of the same family, but of the same litter, or at all events they should have been paired when they were puppies. It is at no time safe to leave the greater part of them loose; and the younger dogs are described as the most dangerous in this way. They will not only at all times kill domestic fowls,—which the Kamtschatkans, therefore, are unable to breed,—and dogs of the smaller species that may chance to be brought to the place, but they have been known to destroy children. While they do not work they are tolerably fat, and have usually an allowance of half a dried salmon, or a portion weighing about two pounds, a day; but when they labour they are worse treated and more stinted than the Siberian horses, and receive only half the quantity of food apportioned to them when at rest; yet they will, under this treatment, perform journeys of three or four weeks' duration with much less repose than the horses require. Nay, they will even, upon a journey of four or five days' duration, work for fourteen or sixteen hours out of the twenty-four without tasting any food whatsoever, and without appearing to suffer any diminution of strength; and the universal opinion seems to be, that the less food they receive on this side of starvation, when travelling, the better.

Five of these dogs will draw a sledge carrying three full-grown persons and sixty pounds weight of luggage. When lightly loaded, such a sledge will travel from thirty to forty versts in a day over bad roads and through the deep snow, while on even roads it will accomplish eighty to one hundred and twenty. And herein lies the inestimable value of the Kamtschatkan dogs, for the horse would be useless in sledging : in the deep snow it would sink ; and it would be unable, on account of its weight, to cross the rivers and streams which are covered only with a thin sheet of ice.

A KAMTSCHATKAN SLEDGE AND TEAM.

But travelling with dogs is by no means easy. Instead of the whip, the driver uses a crooked stick with iron rings, which, by their jingling, supply the leader of the team with the necessary signals. If the dogs show symptoms of relaxing in their efforts, the stick is cast among them to rouse them to greater speed ; and the driver dexterously picks it up again as his sledge shoots by. In a snowstorm they keep their master comfortably warm, and will lie round about him quietly for hours. They are experienced weather-prophets too, for if, when resting, they dig holes in the snow, it is a certain sign of a storm.

The training of these dogs begins at a very early age. Soon after their birth they are placed with their mother in a deep pit, so as to see neither man nor beast ; and after being weaned, they are still condemned to a total exclusion from "the madding crowd." A probation of six months having expired, they are attached to a sledge with older dogs, and being extremely shy, they run at their very fastest. On returning home they undergo another period of pit-life, until they are considered perfectly trained, and capable of performing a long journey. They are then allowed to enjoy their summer freedom. Such a mode of training may render them docile and obedient, but it renders them also gloomy, mistrustful, and ill-tempered.

Siberia, so far as the valley of the Lena is concerned, and even eastward to the Kolima and westward to the Yenisei, is inhabited by the bold and vigorous race of the Jakuts. Their number is computed at about 200,000, and they inhabit the extensive but dreary province of Jakutsk, with a chief town of the same name.

The Jakuts are to a great extent a pastoral people, but as they trade in horses and cattle, and also carry on a brisk fur-trade with the Russians, they have attained a far higher level of civilization than is common among pastoral races. In summer they live in light conical tents ("urossy"), which are fixed upon poles, and covered with birch rind. These they pitch in the open plains and valleys, and then devote themselves to gathering supplies of hay against the coming winter. This is with them a very important labour, for their chief wealth is in their herds of cattle, and to find a sufficient provision for them in the bleak climate of the Lena basin, and on the borders of the Arctic World, is a task of great difficulty. Often, indeed, the supply fails before the return of spring, and the oxen must then be fed upon the young shoots and saplings of the birch and willow.

When winter approaches, the Jakut removes from his tent into a warm, timber-built hut, or *jart*, which assumes the form of a truncated pyramid, and has an exterior covering of turf and clay. Its windows are made of thin sheets of ice; which, as soon as a thaw sets in, are replaced by fish-bladders or paper steeped in oil. The floor is of earth, very rarely boarded, and generally sunk two or three feet below the surface of the ground. The seats and sleeping-berths are arranged along the sides; the hearth, or *tscherwal*, occupies the centre, and its smoke finds an exit through an aperture in the roof. Clothes and weapons are suspended from the walls, and the general appearance of the interior is squalid and disorderly.

Near the jart are stalls for the cows; but when the cold is very intense, they, like the Irishman's pigs, find accommodation indoors. As for the horses, they remain night and day in the open air, though the weather may be so severe that even mercury freezes; and they have no other food than the decayed autumn grass, which they find under the snow.

The capacity of endurance which the Jakut horses exhibit is almost incredible. Like other quadrupeds in the Arctic regions, they change their hair in summer. Traversing, month after month, the dreary wilderness where the only vegetation is a scanty and half-rotten grass, they still retain their strength and energy; and notwithstanding the hard conditions of their lives, they do not age so quickly as our own more carefully-tended steeds. To aim at improving the Jakut horse would be, in the opinion of many travellers, to gild refined gold, and perfume "thrown on the violet." He will continue a steady trot for hours, over roads of which no Englishman can form an idea, and stop to take his rest with no other food but the bark of the larch and willow, or a little hard grass, no covering protecting his foaming sides from the cold, and the temperature down at 40°.

As the horse, so the master. The Jakut is the very personification of hardiness. He seems able to endure anything, and to attempt everything. On the longest winter-journey he carries neither tents nor extra coverings with him, not even one of the large fur-dresses, such as the Siberians generally use. He contents himself, in fact, with his usual dress; in this he generally sleeps in the open air: his bed, a horse-rug stretched upon the snow; his pillow, a wooden saddle. With the same fur jacket which serves him by daytime as a dress, and which he pulls off when he lies down for the night, he defends his back and shoulders, while the front part of his body is turned towards the fire, almost without any covering. He then stops his nose and ears with small pieces of skin, and covers his face so as to leave only a small aperture for breathing; these are all the precautions he takes against the severest cold. Even in Siberia the Jakuts are known as "iron men."

The horse to the Jakut is as valuable and as important as the camel to the Arab or the reindeer to the Lapp. It is not only his steed, which seems incapable of weariness,—his beast of draught, patient under heavy loads,—but its skin provides him with articles of dress; with its hair he makes his fishing-nets; boiled horse meat is his favourite food, and sour mare's milk, or *koumis*, his principal beverage. By mixing this milk with rye flour, or the inner rind of the fir or larch, he makes a thick porridge, which he flavours with berries, or dried fish, or rancid fat.

Before commerce had been diverted into the valley of the Amur, thousands of pack-horses, under the guidance of Jakuts, annually crossed the Stanowoi hills on the way to Ochotsk; a journey of terrible difficulty, which might appal the stoutest nerves. But the Jakut endures the extremes of cold and hunger with a wonderful equanimity. He fears neither the stormy winds, the darkened heavens, the depth of the treacherous morass, nor the darkness and silence of the forest. Nothing appals him but the unseen presence of "Ljeschei," the spirit of the mountain and the forest. The traveller frequently comes upon a fir-crowned hillock, and from the branches of one of the oldest firs sees suspended innumerable tufts of horse-hair. What does it mean? He needs not to inquire, for, lo! his Jakut driver, dismounting from his steed, hastens to pluck a few hairs from his horse's mane, and then, with much reverence, attaches them to the nearest bough, in order to propitiate the terrible Ljeschei. Even Jakuts who have been baptized, and are nominally enrolled among the Christian population, are guilty of this silly bit of superstition; while it is suspected, on good grounds, that they still cherish their belief in Schamanism, and their ancient dread of evil spirits. When we remember, however, the absurd beliefs and vulgar errors still lingering in many parts of our own land, we are unable to pass a very severe verdict on the credulity of the Jakuts.

When on the road they beguile the tedium of the way by singing songs of the most doleful character, corresponding to the habitual melancholy which they seem to have inherited from their forefathers; a melancholy suggested, probably, by the gloom of the landscape, the chilling aspect of the sky, the inclemency of the climate, and the prolonged battle in which their lives are passed. Their songs, not the less, are songs worthy of a bold and intelligent people, and, like the poetry of the Norsemen, are replete with images borrowed from nature. They constantly describe in glowing language the lofty magnificence of the snow-crowned mountains, the starry beauty of the night, the roll and rush of the river, the wail of the wind as it streams through a forest of pines. The Jakut minstrels are mostly improvisatores; and, to secure the favour of the Ljeschei, they will extol the charms of the wilderness over which it rules, as if that wilderness were a portion of Elysium.

The Jakut merchants are remarkable for their enterprise. Their capital is Jakutsk, on the Lena, and thence they extend their operations in all directions. In the rigour of winter they will lead their caravans to Ochotsk, or Kjachta, or Ostrownoje.

Yet the country they traverse is at all times a desert. The mean temperature of the year is only +14°. In November the thermometer sinks to −40°, or 72° below freezing-point. The Yana, at Nishni Kolymsk, freezes early in September; and lower down, where the current is sluggish, loaded horses can cross its frozen surface as early as the middle of August: yet the ice does not melt before June. The sun remains, it is true, about fifty-two days above the horizon; but its light, shrouded by almost continual mists, is attended by but little heat,—and its orb, compressed by refraction into an elliptical form, may be examined by the naked eye without inconvenience.

As the climate, so the vegetation. Dwarfish willow-shrubs, stunted grass, moss, and a few berry-bearing plants compose the flora of the cheerless tundras. There is greater abundance and more variety in the neighbouring and better sheltered valleys of the Aniuj; the poplar, birch, thyme, absinth, and low creeping cedar enliven their slopes; but even in these places Nature is most niggardly of her gifts. Such is not the case, however, with the fauna of Arctic Siberia. The forests are tenanted by numbers of reindeer, elks, bears, foxes, sables, and gray squirrels; while in the low grounds stone foxes make their burrows. With the return of spring come immense flights of swans, geese, and ducks, which build their nests in the most sequestered corners. The sea-coast is frequented by eagles, owls, and gulls; the brushwood by the white ptarmigan; the brooks by hundreds of little snipes. Even the songs of the finch are not wanting in spring, nor is the thrush wholly silent in autumn.

Summing up the details recorded by Admiral Wrangell, a recent writer draws an impressive picture of the mode of life of the people of this desolate waste, and observes: "All denotes that here the limits of the habitable earth are passed; and one asks with astonishment, What could induce human beings to take up their abode in so comfortless a region?"

The chief resource of the Sullaheris of the River Aniuj is, he says, the reindeer chase,—the success of which mainly determines whether famine or some degree of plenty is to be their lot during the coming winter. The passage of the reindeer takes place twice a year: in spring, when the mosquito-swarms drive them to the sea-coast, where they feed on the moss of the tundra; and in autumn, when the increasing cold forces them to retire inland. The spring migration, which begins about the middle of May, is not very profitable; partly because the animals are then in poor condition, and partly because it is more difficult to kill them as they dash across the frozen rivers. The chief hunting takes place in August and September, when the herds, each numbering several thousand deer, return to the forests. They invariably cross the river at a particular spot, where a flat sandy bank enables them to land with comparative ease; and here they close up their ranks, as it were, under the guidance of the stalwart veterans of the herd.

After a brief pause of hesitation the herd plunge into the waters, and in a few minutes the surface of the river seems alive with swimming reindeer. Now is the hunter's time; and out from his concealment in the reedy creek he darts in his little boat, wounding as many animals as he can. While he and his comrades are thus engaged, they run some risk of being capsized in the turmoil, for the bucks gallantly defend themselves with horns, and teeth, and hind legs, while the roes usually attempt to spring with their fore feet upon the gunwale of the boat. If the hunter should be overset, his sole chance of safety is to cling to a strong animal, which will carry him securely across the stream. Such an accident, however, is of rare occurrence. A good hunter will kill a hundred reindeer, or even more, in half an hour. Meantime, the other boats seize the slaughtered animals, which become the property of their crews; while those that are merely wounded and swim ashore belong to the hunters, who, in the midst of the uproar, when all their strength is tasked to the uttermost, so aim their strokes as only to wound severely the larger animals. The noise of the horns striking against each other, the "incarnadined" waters, the shouts of the hunters, the cries of pain, rage, and alarm of the struggling animals, all form a scene which, once seen, is not easily forgotten.

While the men of Kolymsk are thus engaged during the brief summer-time in hunting, fishing, and hay-making, the women wander over the country, and climb the sides of the mountains, for the purpose of gathering edible roots, aromatic herbs, and various kinds of berries—though the last do not ripen every year. The berry-plucking season at Kolymsk, like the vintage in France or Italy, is a season of mirth, a holiday interval in a hard and laborious life. The young women and girls form large parties, and spend whole days and nights in the open air. When the berries are collected, cold water is poured over them, and they are preserved in a frozen state as an addition to the scanty winter fare. We are told that "social parties" are not unknown at Kolymsk, and probably afford as much or as little entertainment there as in more favoured and more civilized communities. The staple luxury is a deluge of weak tea—very weak, for the aromatic leaves which cheer but not inebriate are very dear at Kolymsk; and as sugar is also a costly article, every guest takes a lump of candy in his mouth, lets the tea which he sips flow by, and then replaces it upon the saucer. It would be considered a breach of courtesy if he consumed the entire lump, which thus is made to do duty at more than one soiree. Next to tea, but not less esteemed, the principal requisite for a Kolymsk entertainment is brandy.

Another important Siberian people are the Tungusi, who spread from the basins of the Upper, Middle, and Lower Tunguska to the western shores of the Sea of Ochotsk, and from the Chinese frontiers and the Baikal to the Polar Ocean. Their number does not exceed thirty thousand. According to their avocations, and the domestic animals which constitute their wealth, they are known as the Reindeer, Horse, Dog, Forest, and River Tungusi. Those who keep or rear horses and cattle are but a few; the majority depend on the reindeer. The condition of all is deplorably wretched. The Tungusi has no resource but fishing or hunting. When the rivers are frozen, he withdraws into the forest. Here his misery is so great and his need so extreme that he frequently becomes a cannibal, and attacks the wives and children of his more fortunate countrymen. In happier circumstances he is remarkable for the readiness of his wit, the vivacity of his manner, and the blithesome carelessness of his disposition. It is asserted, however, that he is both malignant and deceitful. He is vain; and loves to decorate his person with strings and ornaments of glass beads, from his small Tartar cap to the tips of his shoes. When hunting the reindeer, or travelling through the forests, however, he puts on large water-tight boots, or *sari*, well greased with fat; and he carries, on these occasions, a small axe, a kettle, a leathern wallet containing some dried fish, and a short gun, or a bow and a sling. He is always accompanied by his faithful dog.

"With the assistance of his long and narrow snow-shoes, he flies over the dazzling plain; and protects his eyes, like the Jakut, with a net made of black horse-hair. He never hesitates to attack the bear single-handed, and generally masters him. The nomad Tungusi naturally requires a movable dwelling. His tent is covered with leather, or large pieces of pliable bark, which are easily rolled up, and transported from place to place. The *jart* of the sedentary Tungusi resembles that of the Jakut, and is so small that it can be very quickly and thoroughly warmed by a fire kindled on the stone hearth in the centre. In his food the Tungusi is by no means dainty. One of his favourite dishes consists of the contents of a reindeer's stomach mixed with wild berries, and spread out in thin cakes on the rind of trees, to be dried in the air or in the sun. Those who have settled on the Wilnj and in the neighbourhood of Nertschinsk likewise consume large quantities of birch tea, which they boil with fat and berries into a thick porridge; and this unwholesome food adds, no doubt, to the yellowness of their complexion."

We shall now, and lastly, take a glance at the Tchuktche (or Tuski), who inhabit the north-eastern point of Asia, with the ice-covered waters of the Polar Sea on one side, and those of Behring Sea on the other. Their land is but seldom visited; all, however, who have ventured thither agree in describing it as one of the most melancholy regions of the earth. The soil is barren, and half-frozen, yielding no other vegetation than mosses and lichens, the vaccinium, and the dwarf birch and willow,—except in the low grounds, where the reedy marshes are frequented in the summer by geese, and swans, and ducks, and wading-birds. The climate is so rigorous that one wonders man can make up his mind to endure it. There is no summer earlier than the 20th of July; and on the 20th of August the shadow of winter comes upon the earth. Animal life, however, if not very varied, is abundant: walruses, sea-lions, and seals inhabit the coasts; and the reindeer, the wolf, the argali, and the Arctic fox are found in the interior.

The Tchuktche are an enterprising people, and fond of independence. Unlike their neighbours, the Koriaks, they have always maintained their freedom against the encroachments of Russia. They are active and spirited traders. In skin-covered boats they cross Behring Straits, and barter furs and walrus-teeth with the natives of America. In long caravans, their sledges drawn by reindeer, they repair to the great fair of Ostrownoje, and carry on a vigorous commerce with the Russian merchants. In their train follow sledges laden with supplies of lichen and moss for the reindeer, as in their wanderings, however circuitous these may be, they are compelled to traverse broad spaces of stony desert, where even these abstemious animals can obtain no food. As their movements are regulated by the necessities of their herds, they occupy five or six months in a journey which, in a straight line, would not exceed a thousand versts in length; they are almost always migrating from place to place, yet, as they invariably carry their dwellings with them, they never leave home. A caravan generally consists of fifty or sixty families; and as soon as one fair is at an end, they depart to make their preparations for the next.

The great staple of the trade at Ostrownoje is tobacco. To secure a small supply of the narcotic which forms the sole luxury of their dreary lives, the Eskimos of North America, extending from the Icy Cape to Bristol Bay, send their articles of barter from hand to hand as far as the Gwosdus Islands in Behring Strait, where the Tchuktche purchase them with tobacco bought at Ostrownoje. Thus, in the icy regions of the extreme north, tobacco is the source and support of considerable commerce; and the narcotic weed which Raleigh and his contemporaries introduced from America into Europe, and which from Europe made its way into Asia, is exported from Asia for the use of American tribes.

The balance of trade, however, seems entirely against the latter. We are told that the skins which a Tchuktche purchases of an Eskimo for half a pood (eighteen pounds) of tobacco-leaves, he sells to the Russian for two poods (seventy-two pounds); and these skins, costing the Russian about one hundred and sixty roubles, the latter sells at Jakutsk for two hundred and sixty, and at St. Petersburg for upwards of five hundred roubles.

The furs sold at Ostrownoje are chiefly those of stone foxes, black and silver-gray foxes, gluttons, lynxes, otters, beavers, and martens. Other products brought thither by the Tchuktche are bear-skins, walrus-teeth, and thongs, sledge-runners (made of whale ribs), and dresses of reindeer-skin. The Russians, besides tobacco, dispose of kettles, axes, knives, guns, tea, and sugar.

A visit to the family of a Tchuktche chief is thus described by one of Admiral Wrangell's companions:—

We entered the outer tent, or *namet*, consisting of tanned reindeer-skins outstretched on a slender framework. An opening at the top to give egress to the smoke, and a kettle on the hearth in the centre, showed that antechamber and kitchen were here harmoniously blended into one. But where might be the inmates? Most probably in that large sack made of the finest skins of reindeer calves, which occupied, near the kettle, the centre of the *namet*. To penetrate into this "sanctum sanctorum" of the Tchuktch household, we raised the loose flap which served as a door, crept on all fours through the opening, cautiously refastened the flap by tucking it under the floor-skin, and found ourselves in the *polog*—that is, the reception or withdrawing-room. A snug box, no doubt, for a cold climate, but rather low, as we were unable to stand upright in it; nor was it quite so well ventilated as a sanitary commissioner would require, as it had positively no opening for light or air. A suffocating smoke met us on entering: we rubbed our eyes; and when they had at length got accustomed to the pungent atmosphere, we perceived, by the gloomy light of a train-oil lamp, the worthy family sitting on the floor in a state of almost complete nudity. Without being in the least embarrassed, Madame Leütt and her daughter received us in their primitive costume; but to show us that the Tchuktche knew how to receive company, and to do honour to their guests, they immediately inserted strings of glass beads in their hair.

Their hospitality equalled their politeness; for, instead of a cold reception, a hot dish of boiled reindeer flesh, copiously irrigated with rancid train-oil by the experienced hand of the mistress of the household, was soon after smoking before them. The culinary taste of the Russians, however, could not appreciate this work of art, and the Leütt family were left to do justice to it unaided.

The Tchuktche are polygamous. Their women are regarded as slaves, but are not badly treated. Most of the Tchuktche have been baptized, but they cling in secret to their heathen creed, and own the power of the shamans, or necromancers. They form two great divisions: the reindeer, or wandering Tchuktche, who call themselves Tennygk; and the stationary Tchuktche, or Oukilon, who exhibit affinities with the Eskimos, and subsist by hunting the whale, the walrus, and the seal. The Oukilon are supposed to number 10,000, and the Tennygk about 20,000.

CHAPTER X.

HISTORICAL SKETCH OF ARCTIC DISCOVERY.

IN the reign of Henry VIII., Dr. Robert Thorne declared that "if he had facultie to his will, the first thing he would understande, even to attempt, would be if our seas northwarde be navigable to the Pole or no." And it is said that the king, at his instigation, "sent two fair ships, well-manned and victualled, having in them divers cunning men, to seek strange regions; and so they set forth out of the Thames, the 20th day of May, in the nineteenth year of his reign, which was the year of our Lord 1527." Of the details of this expedition, however, we have no record, except that one of the vessels was wrecked on the coast of Newfoundland.

In 1536, a second Arctic voyage was undertaken by a London gentleman, named Hore, accompanied by thirty members of the Inns of Law, and about the same number of adventurers of a lower estate. They reached Newfoundland, which, according to some authorities, was discovered by Sebastian Cabot in 1496, and here they suffered terrible distress; in the extremity of their need being reduced to cannibalism. After the deaths of a great portion of the crew, the survivors captured by surprise a French vessel which had arrived on the coast, and navigated her in safety to England.

But the true history of Arctic Discovery dates, as Mr. Markham observes, from the day when the veteran navigator, Sebastian Cabot, explained to young Edward VI. the phenomena of the variation of the needle. On the same day the aged sailor received a pension; and immediately afterwards three discovery-ships were fitted out by the Muscovy Company under his direction. Sir Hugh Willoughby was appointed to their command, with Richard Chancellor in the *Edward Bonadventure* as his second. The latter, soon after quitting England, was separated from the squadron, and sailing in a northerly direction, gained at last a spacious harbour on the Muscovy coast. Sir Hugh's ship, and her companion, the *Bona Confidentia*, were cast away on a desolate part of the Lapland coast, at the mouth of the river Arzina. They entered the river on September 18, 1563, and remained there for a week; and "seeing the year far spent, and also very evil weather, as frost, snow, and hail, as though it had been the deep of winter, they thought it best to winter there." But as day followed day, and week followed week, in those grim solitudes of ice and snow, the brave adventurers perished one by one; and many months afterwards their bleached bones were discovered by some Russian fishermen.

In the spring of 1556, Stephen Burrough, afterwards chief pilot of England, fitted out the "Search-thrift" pinnace, and sailed away for the remote north. He discovered the strait leading into the sea of Kara, between Novaia Zemlaia and the island Waigatz; but he made up his mind to return, because, first, of the north winds, which blew continually; second, "the great and terrible abundance of ice which we saw with our eyes;" and third, because the nights waxed dark. He arrived at Archangel on September 11, wintered there, and returned to England in the following year.

Twenty years later, on a bright May morning, Queen Elizabeth waved a farewell to Martin Frobisher and his gallant company, as they dropped down the Thames in two small barks, the *Gabriel* and the *Michael*, each of thirty tons, together with a pinnace of ten tons. They gained the shores of Friesland on the 11th of July; and sailing to the south-west, reached Labrador. Then, striking northward, they discovered "a great gut, bay, or passage," which they named Frobisher Strait (lat. 63° 8′ N.), and fell into the error of supposing that it connected the Atlantic Ocean with the Pacific. Here they came into contact with some Eskimos; and Frobisher describes them as "strange infidels, whose like was never seen, read, nor heard of before: with long black hair, broad faces and flat noses, and tawny in colour, wearing seal-skins, the women marked in the face with blue streaks down the cheeks, and round about the eyes."

Frobisher's discoveries produced so great an impression on the public mind, that in the following year he was placed at the head of a larger expedition, in the hope that he would throw open to English enterprise the wealth of "far Cathay." About the end of May 1577, he sailed from Gravesend with the *Ayde* of one hundred tons, the *Gabriel* of thirty, and the *Michael* of thirty, carrying crews of ninety men in all, besides about thirty merchants, miners, refiners, and artisans. He returned in September with two hundred tons of what was supposed to be gold ore, and met with a warm reception. It was considered almost certain that he had fallen in with some portion of the Indian coast, and Queen Elizabeth, naming it *Meta Incognita*, resolved to establish there a colony. For this purpose, Frobisher was dispatched with fifteen well-equipped ships, three of which were to remain for a twelvemonth at the new settlement, while the others, taking on board a cargo of the precious ore, were to return to England.

In the third week of June Frobisher arrived at Friesland, of which he took possession in the queen's name. Steering for Frobisher Strait, he found its entrance blocked up with colossal icebergs; and the bark *Dennis*, which carried the wooden houses and stores for the colony, coming in collision with one of these, unfortunately sank. Then, in a great storm, the fleet was scattered far and wide,—some of the vessels drifting out to sea, some being driven into the strait; and when most of them rejoined their admiral, it was found they had suffered so severely that no help remained but to abandon the project of a colony. They collected fresh supplies of ore, however, and then made their way back to England as best they could. Here they were met with the unwelcome intelligence that the supposed gold ore contained no gold at all, and was, in truth, mere dross and refuse.

The dream of a northern passage to Cathay was not to be dissipated, however, by an occasional misadventure. Even a man of the keen intellect of Sir Humphrey Gilbert felt persuaded that through the northern seas lay the shortest route to the treasures of the East; and having obtained from Queen Elizabeth a patent authorizing him to undertake north-western discoveries, and to acquire possession of any lands not inhabited or colonized by Christian princes or their subjects, he equipped, in 1583, with the help of his friends, a squadron of five small ships, and sailed from England full of bright visions and sanguine anticipations. On board his fleet were smiths, and carpenters, and shipwrights, and masons, and refiners, and "mineral men;" not to speak of one Stephen Parmenio, a learned Hungarian, who was bound to chronicle in sonorous Latin all "gests and things worthy of remembrance."

Sir Humphrey formed a settlement at Newfoundland; and then, embarking on board the *Squirrel*, a little pinnace of ten tons burden, and taking with him the *Golden Hind* and the *Delight*, he proceeded on a voyage of exploration. Unhappily, the *Delight* ran ashore on the shoals near Sable Land, and all her crew except twelve men, and all her stores, were lost. The disaster determined Sir Humphrey to return to England; and his companions implored him to embark on board the *Golden Hind*, representing that the *Squirrel* was unfit for so long a voyage. "I will not forsake," replied the chivalrous adventurer, "the brave and free companions with whom I have undergone so many storms and perils." Soon after passing the Azores, they were overtaken by a terrible tempest, in which the tiny pinnace was tossed about by the waves like a straw. The *Golden Hind* kept as near her as the rolling billows permitted; and her captain has left on record that he could see Sir Humphrey sitting calmly in the stern reading a book. He was heard to exclaim—"Courage, my lads; we are as near heaven by sea as by land!" Then night came on, with its shadows and its silence, and next morning it was perceived that the pinnace and her gallant freight had gone to swell the sum of the irrecoverable treasures of the deep.

THE LOSS OF THE "SQUIRREL."

But neither Frobisher's mishap nor Sir Humphrey Gilbert's melancholy fate could check that current of English enterprise which had set in for the North. There was an irresistible attraction in these remote northern seas and distant mist-shrouded lands, with all their possibilities of wealth and glory; and Arctic Discovery had already begun to exercise on the mind of the English people that singular fascination which the course of centuries has not weakened, which endures even to the present day. So, in 1585, Sir Adrian Gilbert and some other gentlemen of Devonshire raised funds sufficient to fit out a couple of vessels—the *Sunshine* of fifty, and the *Moonshine* of thirty-five tons—for the great work of discovery; and they gave the command to a veteran mariner and capable navigator, Captain John Davis, a countryman, or county-man, of their own. Towards the end of July he reached the west coast of Greenland, and its cheerless aspect induced him to christen it the "Land of Desolation." His intercourse with the Eskimos, however, was of the friendliest character. Standing away to the north-west, he discovered and crossed the strait which still bears his name; and to the headland on its western coast he gave the name of Cape Walsingham. Having thus opened up, though unwittingly, the great highway to the Polar Sea, he sailed for England, where he arrived on the 20th of September.

In his second voyage, in 1586, when, in addition to the *Sunshine* and the *Moonshine*, he had with him the *Mermaid* of one hundred and twenty tons, and the *North Star* pinnace of ten, he retraced his route of the previous year. The *Sunshine* and the *North Star*, however, he employed in cruising along the east coast of Greenland; and they ascended, it is said, as high as lat. 80° N.

Davis in his third voyage pushed further to the north, reaching as far as the bold promontory which he named Cape Sanderson. He also crossed the great channel afterwards known as Hudson Bay.

The next Englishman who ventured into the frozen seas was one Captain Waymouth, in 1602; but he added nothing to the scanty information already acquired. An Englishman, James Hall, was the chief pilot of an expedition fitted out in 1605 by the King of Denmark, which explored some portion of the Greenland coast. He made three successive voyages; but while exhibiting his own courage and resolution, he contributed nothing to the stores of geographical knowledge.

We now arrive at a name which deservedly ranks among the foremost of Arctic explorers—that of Henry Hudson. He contributed more to our acquaintance with the Polar seas than any one who had preceded him, and few of his successors have surpassed him in the extent and thoroughness of his researches.

He first appears, says Mr. Markham, fitting out a little cock-boat for the Muscovy Company, called the *Hopewell* (of eighty tons), to discover a passage by the North Pole. On the 1st of May 1607 he sailed from Greenwich. "When we consider the means with which he was provided for the achievement of this great discovery, we are astonished at the fearless audacity of the attempt. Here was a crew of twelve men and a boy, in a wretched little craft of eighty tons, coolly talking of sailing right across the Pole to Japan, and actually making as careful and judicious a trial of the possibility of doing so as has ever been effected by the best equipped modern expeditions......Imagine this bold seaman sailing from Gravesend, bound for the North Pole, in a craft about the size of one of the smallest of modern collier brigs. We can form a good idea of her general appearance, because three such vessels are delineated on the chart drawn by Hudson himself. The *Hopewell* was more like an old Surat buggalow than anything else that now sails the seas, with high stern, and low pointed bow; she had no head-sails on her bowsprit, but, to make up for this, the foremast was stepped chock forward. There was a cabin under the high and narrow poop, where Hudson and his little son were accommodated; and the crew were crowded forward."

SHIP OF THE SEVENTEENTH CENTURY.

Hudson first sighted land beyond the Arctic Circle in lat. 70°. It was the cold, grim coast of East Greenland. Three degrees further north a chain of lofty peaks, all bare of snow, rose upon the horizon, and Hudson's men noted that the temperature daily increased in mildness. Steering to the north-east, the great navigator arrived off the shores of Spitzbergen, where some of his men landed and picked up various fragments of whalebone, horns of deer, walrus-teeth, and relics of other animals. To the north-west point of Spitzbergen he gave the name which it still bears—Hakluyt's Headland. At one time he found himself as far north as 81°; and it seems probable that he discovered the Seven Islands: he remarked that the sea was in some places green, in others blue; and he says, "Our green sea we found to be freest from ice, and our azure-blue sea to be our icy sea;" an observation not confirmed by later navigators. The greenness was probably due to the presence of minute organisms.

SCENERY OF JAN MAYEN.

Having completed a survey of the west coast of Spitzbergen, he resolved on sailing round the north end of Greenland, which he supposed to be an island, and returning to England by Davis Strait. With this view he again examined the sea between Spitzbergen and Greenland, but from the strong ice-blink along the northern horizon felt convinced that there was no passage in that direction. After sighting Spitzbergen, therefore, he determined to return to England; and on his homeward voyage discovered an island in lat. 71° N., which he named Hudson Sutches, and which has since been improperly named Jan Mayen. The *Hopewell* arrived in the Thames on the 15th of September.

The results of this voyage, says Mr. Markham, were very important, both in a geographical and a commercial point of view. Hudson had discovered a portion of the east coast of Greenland; he had examined the edge of the ice between Greenland and Spitzbergen twice—in June and in the end of July; and he had sailed to the northward of Spitzbergen until he was stopped by the ice, reaching almost as high a latitude as Scoresby in 1806, which was 81° 12′ 42″ N. Hudson's highest latitude by observation was 80° 23′, but he sailed for two more days in a north-easterly direction. The practical consequence of his voyage was that his account of the quantities of whales and sea-horses in the Spitzbergen seas led to the establishment of a rich and prosperous fishery, which continued to flourish for two centuries.

In the following year Hudson made a second voyage, in the hope of discovering a north-eastern passage to China between Spitzbergen and Novaia Zemlaia. He exhibited his characteristic resolution, and forced his way to the very gate of the unknown region, which is still closed against human enterprise by an impenetrable barrier of ice; but all his efforts proved in vain, and he returned to Gravesend on the 26th of August.

In 1610, in a vessel of fifty-five tons, he once more entered the Polar seas, and gained the extreme point of Labrador, which he named Cape Wolstenholm. Here burst upon him the view of that magnificent sea which has since been associated with his name; and there can be no doubt that his enterprise would have anticipated the discoveries of later navigators, but for the mutiny which broke out among his crew, and eventually led to his being sent adrift, with nine faithful companions, in a small open boat. He was never again heard of.

The spirit of commercial enterprise and the love of maritime adventure were still strong enough in England to induce the equipment of further expeditions. In 1612 sailed Captain Button,—who discovered a stream, and named it Nelson River; where, at a later date, the Hudson Bay Company planted their first settlement. Here he wintered. In April 1613, on the breaking up of the ice, he resumed his work of exploration, and discovered, in lat. 65°, an island group, which he named Manuel, now known as Mansfield, Islands. Then he bore away for England, arriving in the Thames early in September.

Robert Bylot and William Baffin undertook a voyage in 1615. The latter had had some previous experience of Arctic navigation, which he turned to advantage in 1616, when he accompanied Bylot on a second expedition. Their ship, the *Discovery*, of fifty-five tons, reached Cape Hope Sanderson, the furthest point attained by Davis, on the 30th of May; and after meeting with some obstruction from the ice, proceeded northwards to 72° 45′, where she dropped anchor for awhile among the Women's Islands. Baffin kept to the north until he found ice in 74° 15′ N., and he then ascended Melville Bay, touching the head of the great basin now known by his name, and sailing down its western coast. He arrived in Dover Roads on the 30th of August, after a brilliantly successful voyage, which had opened up the principal north-west channels into the Arctic Sea.

It is necessary here to interpolate a few remarks in explanation of the difficulties which beset the Baffin Bay route of Arctic exploration. Geographers assert, and the assertion seems confirmed by the experience of navigators, that a surface-current is constantly flowing down this bay, and carrying great fleets of icebergs and shoals of ice-floes into the Atlantic from its southern channels—Lancaster, Jones, and Smith Sounds. Hence, at the head of the bay there exists a considerable open and navigable expanse, which extends for some distance up Lancaster and Smith Sounds during the summer and early winter, and is known as the "North Water." But between this open expanse and Davis Strait lies an immense mass of ice, averaging from one hundred and seventy to two hundred miles in width, and blocking up the centre of Baffin Bay, so as to interrupt the approach to the north-west end. This is known as the "middle pack," and consists of some ancient floe-pieces of great thickness, which may have been brought down from a distant part of the Arctic seas; of a wide extent of ice accumulated during each winter, about six or eight feet in thickness; and of the grand and gigantic icebergs which are so characteristic a feature of the Melville Bay scenery. A very large quantity of this pack is destroyed in each succeeding summer by the thaws, or by the swell and warm temperature of the Atlantic as the ice drifts southward.

It is remarked of the Baffin Bay ice, that it is much lighter than that found in the Spitzbergen seas. The latter often occurs in single sheets, solid, transparent, and from twenty to thirty, and even forty, feet in thickness. In Baffin Bay the average thickness of the floes does not exceed five or six feet, and eight or ten feet is of very rare occurrence.

From Baffin's voyage, in 1616, until 1817, no attempt was made to force this "middle pack" and enter the North Water; but now the voyage is made every year, and three routes have been opened up. The first is called the "North-about Passage," and lies along the Greenland coast; the second, or "Middle Passage," only possible late in the season, is by entering the drift-ice in the centre of the bay; and the third, or "Southern Passage," also only possible late in the season, along the west side of Baffin Bay. Once in the North Water, whichever route be attempted, all obstacles to an exploration of the unknown region may be considered at an end. From Cape York to Smith Sound the sea is always navigable in the summer months.

It will thus be seen that the great highways to the Pole were discovered by William Baffin.

Our limits compel us to pass over the voyages of Stephen Bennet (1603–1610), Jonas Poole (1610–1613), and Captain Luke Fox (1631). In 1631 the merchants of Bristol despatched Captain Thomas James, but he made no additions to the discoveries of his predecessors. And then for nearly two centuries England abandoned her efforts to open up a communication between the Atlantic and the Pacific.

In 1818, however, the question of the existence of a North-West Passage once more occupied the public mind; and the British Government accordingly fitted out an exploring expedition, the *Isabella* and the *Alexander*, under the command of Captain Ross and Lieutenant Parry.

They sailed from England on the 18th of April, reached the southern edge of the Baffin Bay ice on the 2nd of July, and, after a detention of thirty-eight days, reached the North Water on August 8th. The capes on each side of the mouth of Smith Sound, Ross named after his two ships; and having accomplished this much, he affirmed that he saw land against the horizon at a distance of eight leagues, and then retraced his course, and sailed for England.

The British Government, however, refused to be discouraged by the failure of an expedition which had obviously been conducted with an entire absence of vigour and enterprise. They therefore equipped the *Hecla* and the *Griper*, and gave the command to Lieutenant Parry; who sailed from the Thames on the 5th of May 1819, and on the 15th of June sighted Cape Farewell. Striking northward, up Davis Strait and Baffin Bay, he found himself checked by the ice-barrier in lat. 73° N. A man of dauntless resolution, he came to the determination of forcing a passage at all hazards; and in seven days, by the exercise of a strong will, great sagacity, and first-rate seamanship, he succeeded in carrying his ships through the pack of ice, which measured eighty miles in breadth.

He was then able to enter Sir James Lancaster Sound; and up this noble inlet he proceeded with a fair wind, hopeful of entering the great Polar Sea. But after advancing a considerable distance, he was once more met by the frozen powers of the North, and this time he was forced to own himself vanquished. He accordingly returned towards the south, discovering

Barrow Strait; and, more to the westward, an inlet which has since figured conspicuously in Arctic voyages—Wellington Channel. Bathurst Island he also added to the map; and afterwards he came in sight of Melville Island. On the 4th of September he attained the meridian of 110° W. long., and thus became entitled to the Parliamentary grant of £5000. A convenient harbour in the vicinity was named the "Bay of the *Hecla* and the *Griper*," and here Lieutenant Parry resolved upon passing the winter.

In the following spring he resumed his adventurous course, and completed a very careful survey of the shores of Baffin Sea; after which he repaired to England, and reached the Thames in safety, with his crews in good health, and his ships in excellent condition, about the middle of November 1820.

THE "HECLA" AND "FURY" WINTERING AT WINTER ISLAND.

Having done so much and so well, it was natural that Captain Parry should again be selected for employment in the Arctic seas in the following year. He hoisted his flag in his old ship, the *Hecla*, and was accompanied by the *Fury*; both vessels being equipped in the most liberal manner. He sailed from the Nore on the 8th of May 1821; he returned to the Shetland Islands on the 10th of October 1823. In the interval, a period of seven-and-twenty months, he discovered the Duke of York Bay, the numerous inlets which break up the northern coast-line of the American continent, Winter Island, the islands of Anatoak and Ooght, the Strait of the *Fury* and *Hecla*, Melville Peninsula, and Cockburn Island. During their winter sojourn on Winter Island, the English crews were surprised by a visit from a party of Eskimos, whose settlement Captain Parry visited in his turn. He found it an establishment of five huts, with canoes, sledges, dogs, and above sixty men, women, and children, as regularly, and, to all appearance, as permanently fixed as if they had occupied the same spot the whole winter. "If the first view," says Parry, "of the exterior of this little village was such as to create astonishment, that feeling was in no small degree heightened on accepting the invitation soon given us to enter these extraordinary houses, in the construction of which we observed that not a single material was used but snow and ice. After creeping through two low passages, having each its arched doorway, we came to a small circular apartment, of which the roof was a perfect arched dome. From this three doorways, also arched, and of larger dimensions than the outward ones, led into as many inhabited apartments—one on each side, and the other facing us as we entered. The interior of these presented a scene no less novel than interesting: the women were seated on the beds at the sides of the huts, each having her little fireplace or lamp, with all her domestic utensils about her. The children crept behind their mothers, and the dogs shrank past us in dismay. The construction of this inhabited part of the hut was similar to that of the outer apartment,—being a dome, formed by separate blocks of snow laid with great regularity and no small art, each being cut into the shape requisite to form a substantial arch, from seven to eight feet high in the centre, and having no support whatever but what this principle of building supplies. Sufficient light was admitted into these curious edifices by a circular window of ice, neatly fitted into the roof of each apartment."

THE "FURY" ABANDONED BY PARRY—1825.

In 1824–25 Captain Parry undertook a third voyage, but with less than his usual success. The *Fury* was driven ashore by the pressure of the pack-ice, and so damaged, that Parry found it needful to abandon her, and remove her crew and stores to the *Hecla*.

Sir John Parry's fourth and last expedition, in 1827, was characterized by his bold attempt to cross the icy sea in light boats and sledges; resorting to the former when his progress was interrupted by pools of water, and to the latter in traversing the unbroken surface of the ice-fields. He was soon compelled, however, to abandon the sledges, on account of the hummocks and irregularities of the ice.

We agree with Mr. Cooley, that voluntarily to undertake the toil and brave the danger of such an expedition, required a zeal little short of enthusiasm. When the travellers reached a water-way, they were obliged to launch their boats and embark. On reaching the opposite side, their boats were then to be dragged, frequently up steep and perilous cliffs, their lading being first removed. By this laborious process, persevered in with little intermission, they contrived to accomplish eight miles in five days. They travelled only during the night, by which means they were less incommoded with snow-blindness; they found the ice more firm and consistent; and had the great advantage of lying down to sleep during the warmer portion of the twenty-four hours. Shortly after sunset they took their breakfast; then they laboured for a few hours before taking their principal meal. A little after midnight, towards sunrise, they halted as if for the night, smoked their pipes, looked over the icy desert in the direction in which the journey was to be resumed; and then, wrapping themselves in their furs, lay down to rest. Advancing as far north as 82° 40′, they were then compelled by the drifting of the snow-fields to retrace their steps. They regained their ships on the 21st of August, and sailed for England.

We must now go back a few years. In May 1819, an overland expedition was despatched to ascertain the exact position of the Coppermine River, to descend it to its mouth, and to explore the coast of the Arctic Sea on either hand. The command was given to Lieutenant Franklin, who was accompanied by Dr. Richardson the naturalist, by Messrs. Hood and Back, two English midshipmen, and two picked seamen. The expedition was spread over a period of two years and a half, and the narrative of what was accomplished and endured by its members reads like a romance. They reached the mouth of the Coppermine, and then launched their little barks on the chill waters of the Polar Sea. With much perseverance, and after encountering some serious obstacles, they made their way along its shores in a westerly direction as far as Point Turnagain, in lat. 68° 30′ N. Between this headland on the east, and Cape Barrow on the west, opens a deep gulf, stretching inland as far as the Arctic Circle. Franklin named it George the Fourth's Coronation Gulf; and describes it as studded with numerous islands, and indented with sounds affording excellent harbours, all of them supplied with small rivers of fresh water, abounding with salmon, trout, and other fish.

Passing over Franklin's after-labours in the great cause of Arctic Discovery, labours which secured him the well-merited reward of knighthood, we come to that last voyage, which helped, as we shall see, to solve the problem of a North-West Passage, but was the cause of one of the saddest chapters in the history of Maritime Enterprise.

It was in the spring of 1845 that Sir John Franklin, in command of the *Erebus* and the *Terror*, with Captain Crozier, an experienced Arctic navigator, as his lieutenant, and at the head of one hundred and thirty-seven picked seamen, brave, resolute, and hardy, once more sailed for the Polar waters.

On the 8th of June he left the Orkneys, and a month later arrived in Baffin Bay. About the end of July some whaling-ships in Melville Bay saw the *Erebus* and *Terror* contending gallantly with the ice which impeded their progress to Lancaster Sound. On the evening of the 26th the ice opened up, and the two discovery-ships sailed away into the north-western seas.

Two years passed, and no news reached England of Franklin and his companions. As day succeeded day, and week followed week, and still no tidings came, men grew anxious, and then alarmed; "expectation darkened into anxiety, anxiety into dread." At last, it was determined to institute a search for the missing heroes. An expedition was sent out under Sir James Ross; another under Sir John Richardson; but neither obtained any information. By many all hope was then abandoned; and the fate of Franklin was regarded as one of those mysteries which the historian in vain attempts to unravel. He and his men had perished; of *that* there could be no reasonable doubt. Yet a few were sanguine enough to believe that they had taken refuge among the Eskimos, or were dragging out a weary existence in some remote wilderness, in expectation of help from home. Franklin's brave and noble wife was one of those who, whatever they feared or hoped, were, at all events, determined not to rest until some accurate information had been gained. And round her gathered the most eminent scientific men of the day, whose influence combined with the general sympathy of the people to encourage the Government in a further effort.

It was in 1850 that the first clue to the position of the *Erebus* and *Terror* was secured in Beechey Island, through the accidental detention there of the searching expeditions of Captains Austin and Penny.

They were bound for Melville Island, but on reaching the entrance of Wellington Channel (August 1850), were met by such immense fields of ice sweeping down it and out of Barrow Strait, that they were glad to seek shelter in a great bay at the eastern end of the channel,—a bay almost bisected, as it were, by Beechey Island. On the 23rd, a boat from Captain Ommaney's ship, the *Assistance*, happened to land on one of the extreme points of the bay; and the crew, in the course of their wanderings, were not a little surprised to discover traces of a former visit from Europeans. Under the lofty cliff of Cape Riley they came upon the groundwork of a tent, scraps of canvas and rope, a quantity of birds' bones and feathers, and a long-handled rake which, apparently, had been used for collecting the rich rare weeds that cover the bottom of the Arctic waters.

That Europeans had been encamped there, was certain, but not a name or record associated the remains with Franklin's expedition. News of the discovery, however, reached Captain Penny, an Aberdeen seaman, who had been employed by the British Admiralty as leader of a separate expedition; and in conjunction with Lieutenant de Haven, of the United States Navy, who was in command of the expedition liberally equipped by Mr. Grinnell, of New York, he resolved to examine the east coast of Wellington Channel with minute care, in the belief that some memorials of Franklin would thus be discovered.

From a point called Cape Spenser, the Americans, on foot, pursued the trail of a sledge up the east side of Wellington Channel, until, at one day's journey beyond Cape Innis, it suddenly ceased, as if the party had there turned back again. A bottle and a piece of *The Times* newspaper were the only relics which fell into the hands of the searchers. Meantime, Captain Penny had anchored his ships under the western point of Beechey Island, and despatched a boat to take up the clue at Cape Riley, and follow it to the eastward, in the event of the traces being those of a party retreating from the ships, supposing them to have been ice-bound in the north-west, to Baffin Bay. This boat-party eventually returned unsuccessf[illegible] but, one afternoon, some men belonging to the *Lady Franklin* asked leave, and obtained it, for the purpose of a ramble over Beechey Island. They sauntered along towards the low projecting portion of the island which extends northward, choosing a convenient spot to cross the huge ridges of ice which lay piled up along the beach; they were seen to mount the acclivity or backbone of the point. In a minute afterwards their friends on board the ships (says Admiral Sherard Osborn) saw the party rush simultaneously towards a dark object, round which they collected, with signs of great excitement. Presently one ran hither, one thither. Feverish with anxiety, those on board knew immediately that some fresh traces had been found, and a general sortie took place to Beechey Island. "Eh, sir," said a gallant Scotch mariner, when relating the discovery--"oh, sir, my heart was in my mouth, and I didna ken I could rin so fast afore."

And what had been found?

A cairn, of a pyramidal form, which had evidently been constructed with much care. The base consisted of a series of preserved-meat tins, filled with gravel and sand; and more tins were so arranged as to taper gradually upwards to the summit of the cairn, in which was planted the fragment of a broken boarding-pike. To all appearance it had been purposely raised for the reception of some documental record, yet nothing could be found in or about the spot, in spite of the most persevering efforts. But presently looking along the northern slope of the island, other strange objects caught the eye. Another rush of anxious excited beings, and they stood before three graves; and many of them brushed away the unwonted tear as they read upon their humble tablets the words *Erebus* and *Terror*.

Captain Austin followed up Captain Penny in his explorations of the Arctic wastes, but no further information was obtained of Franklin's movements. It was impossible to determine whether on his way home he had perished in Baffin Bay; whether he had struck to the north-west by Wellington Channel; or whether he was haply imprisoned in Melville Island.

We have no space, nor is it necessary, to dwell on the records of the various searching expeditions fitted out by the Government, or by Lady Franklin and her friends. It must be noted, however, that one of these, led by Captain (afterwards Sir) Robert M'Clure, succeeded in accomplishing the enterprise in which Franklin perished, and, entering the Northern Ocean by Behring Strait, actually forced its way, through snow and ice, into the Atlantic. The North-West Passage, so long sought, was thus discovered; but the discovery, though interesting and valuable from a geographical point of view, was followed by no commercial results. In truth, it proved that the route along the north-west of the American Continent could never be practicable for ordinary vessels.

It may be asserted that nearly all men had abandoned hope and expectation of ascertaining any exact particulars of the fate of Franklin and his followers, when, towards the close of the autumn of 1854, Dr. Rae, a well-known traveller and Arctic explorer, suddenly appeared in England, bringing with him the most curious evidence of the disasters which had overwhelmed a party that had evidently been travelling from the ice-bound *Erebus* and *Terror* towards the Great Fish River. Dr. Rae had ascertained from some Eskimos with whom he had been travelling that this party numbered forty persons, and that all had died of starvation four years prior to Dr. Rae's visit. The unfortunate "white men" had been first seen on King William's Land; later in the same year their dead bodies had been observed near or about the mouth of the Great Fish River (1850). Dr. Rae brought home numerous pieces of silver plate obtained from the Eskimos, which were marked with the names of officers of the two ships. Lady Franklin was encouraged by this intelligence to urge upon the Government the propriety of despatching an expedition to the points indicated by the Eskimos; but the Government contented themselves with applying to the Hudson Bay Company. The result was an overland expedition in 1855 to the mouth of the Great Fish River, by Mr. Anderson, one of the Company's chief officers. He had no boat with him capable of reaching King William's Land, though it was only sixty miles distant from the point he attained, nor was he accompanied by an Eskimo interpreter. He ascertained, however, that only a portion of the officers and men of the *Erebus* and *Terror* had reached the Great Fish River—some forty of them, very possibly, as Dr. Rae had been informed; these forty, with the three graves upon Beechey Island, still leaving ninety-five persons unaccounted for.

Lady Franklin and her friends continued to press upon Government the need for further inquiry; but finding the responsible ministers unwilling to interfere in what they had come to consider a hopeless enterprise, they contrived, with some help from the public, to purchase and fit out a strongly-built screw-schooner, of which Captain M'Clintock volunteered to take the command.

He sailed from England in the summer of 1857; reached Melville Bay in safety, but was then held fast by the floating ice. The winter, however, came and went without any injury to him and his gallant band; and on the 27th of July 1858, the *Fox* stretched across to Lancaster Sound. On the 11th of August she arrived at Beechey Island, and replenished her diminished stores from the depôts left there by previous expeditions. Then she pushed to the westward, past Cape Hotham and Griffith Island, southward through Sir Robert Peel Channel, and so into Prince Regent Inlet. Having arrived off the eastern entrance of Bellot Strait, she found it blocked up by a wall of ice, and from the 20th of August to the 6th of September she watched for an opportunity of breaking through it. On the 6th she made the passage, but only to find the other end obstructed by an impassable ice-barrier; and, after five fruitless attempts, her captain brought her to anchor for the winter in Port Kennedy, on the northern side of the strait.

When the new year opened, M'Clintock resolved on undertaking sledge excursions in various directions, with the view of obtaining some information of Franklin and his expedition. In one of them, at Cape Victoria, on the west coast of Boothia (lat. 69° 50′ N., long. 96° W.), he ascertained from the natives that, several years previously, a ship had been wrecked off the northern shores of King William's Land; that all her crew landed safely, and set off on a journey to the Great Fish River, where they died. Again: in April, falling in with the same party of Eskimos, they learned further, that besides the ship which had sunk in deep water, another had been driven ashore by the ice. Captain M'Clintock thereupon crossed to Montreal Island, travelled round the estuary of the Great Fish River, and visited Point Ogle and Barrow Island. On May 7, he fell in with an old Eskimo woman, who told him that many of the white men dropped by the way as they made towards the Great Fish River; that some were buried, and some were not. Proceeding in what he conceived to have been the route of the retreating crews, he discovered, near Point Herschel, a bleached skeleton; evidently that of one who had fallen behind the main body, from weakness and fatigue, and had died where he had fallen.

Meanwhile, Lieutenant Hobson, who had started with another sledging party, had made the important discovery of a record, giving a brief account of the Franklin expedition up to the time when the ships were lost. It was found within a cairn constructed on Point Victory, and it set forth the following particulars:—

The *Erebus* and *Terror* spent their first winter at Beechey Island, in the spot discovered by Penny and Austin's expedition; but they had previously explored Wellington Channel as far as 77° N., and passed down again into Barrow Strait, between Cornwallis and Bathurst Land. In 1846 the two ships seem to have sailed through Peel Channel, until caught in the ice off King

William's Land, on the 12th of September. In May 1847, Lieutenant Graham Gore and Mr. des Vœux landed, and erected a cairn a few miles south of Point Victory, and deposited in it a

DISCOVERY OF THE CAIRN CONTAINING SIR JOHN FRANKLIN'S PAPERS.

document which stated that, on that day, all were well, with Sir J. Franklin in command. Within a month, however, that illustrious navigator died (June 11), and thus was spared the

RELICS OF THE FRANKLIN EXPEDITION BROUGHT BACK TO ENGLAND.

terrible trials which afflicted his followers. The ice did not move, and the winter of 1847–48 closed in upon them. It proved fatal to nine officers and fifteen men. On April 22, 1848, the two ships, which had been imprisoned for upwards of nineteen months, were deserted, and the officers and crews, one hundred and five in number, under the command of Captains Crozier and Fitzjames, started for the Great Fish River.

At the cairn and all about it lay a great quantity of clothing and other articles, which the sufferers had found from experience of three days to be a heavier weight than their enfeebled strength was able to drag.

From this point to a spot about midway between Point Victory and Point Herschel nothing of much importance was discovered, and the skeletons as well as relics were deeply embedded in snow. At this midway station, however, the top of a piece of wood projecting out of the snow was seen by Lieutenant Hobson, and on digging round it a boat was discovered. It stood on a very heavy sledge, and within it were a couple of skeletons. The one in the bottom of the stern-

DISCOVERY OF ONE OF THE BOATS OF THE FRANKLIN EXPEDITION.

sheets was covered with a great quantity of thrown-off clothing; the other, in the bows, seemed to have been that of some poor fellow who had crept there to look out, and in that position fallen into his last sleep. A couple of guns, loaded and ready cocked, stood upright to hand, as if they had been prepared for use against wild animals. Around this boat was another accumulation of cast-off articles; and it was the belief of M'Clintock that the party in charge of her were returning to the ships, as if they discovered their strength unequal to the terrible journey before them. It may be assumed, however, that the stronger portion of the crews still pushed on with another boat, and that some reached Montreal Island and ascended the Great Fish River.

The point, says Sherard Osborn, at which the fatal imprisonment of the *Erebus* and *Terror* in 1846 took place, was only ninety miles from the spot reached by Dease and Simpson in their boats in 1838–39, coming from the east. Ninety miles more of open water, and Franklin and his gallant crew would have not only won the prize they sought, but reached their homes to wear their well-earned honours. "It was not to be so. Let us bow in humility and awe to the inscrutable decrees of that Providence who ruled it otherwise. They were to discover the great highway between the Pacific and the Atlantic. It was given them to win for their country a discovery for which she had risked her sons and lavishly spent her wealth through many centuries; but they were to die in accomplishing their last great earthly task: and, still more strange, but for the energy and devotion of the wife of their chief and leader, it would in all probability never have been known that they were indeed the *First Discoverers of the North-West Passage.*

We have thought it for the convenience of our readers to set before them an uninterrupted narrative of the exertions made to ascertain the fate of Franklin and his companions by English seamen under English influence; but we must now return to 1853, to chronicle the American expedition under Dr. Kane—which did not, indeed, succeed in its primary object, but made some remarkable additions to our knowledge of the Polar Regions.

Dr. Elisha Kane sailed from Boston in 1853, in command of the *Advance*, with a crew of seventeen officers and men, to whom two Greenlanders were subsequently added.

On the 7th of August he passed the two great headlands which guard the entrance of Smith Sound,—Cape Isabella and Cape Alexander, discovered and named in the preceding year by Captain Inglefield,—and after a voyage of equal difficulty and danger reached Rensselaer Bay on the east coast of the sound, where he passed the winter. A few extracts from his diary will show under what conditions, and in what circumstances, Kane and his followers passed the long and dreary winter months:—

"*October 28th.*—The moon has reached her greatest northern declination of about 25° 35'. She is a glorious object; sweeping around the heavens, at the lowest part of her curve, she is still 14° above the horizon. For eight days she has been making her circuit with nearly unvarying brightness. It is one of those sparkling nights that bring back the memory of sleigh-bells and songs, and glad communings of hearts in lands that are far away.

"*November 7th.*—The darkness is coming on with insidious steadiness, and its advances can only be perceived by comparing one day with its fellow of some time back. We still read the thermometer at noonday without a light, and the black masses of the hills are plain for about five hours, with their glaring patches of snow; but all the rest is darkness. The stars of the sixth magnitude shine out at noonday. Except upon the island of Spitzbergen, which has the advantages of an insular climate, and tempered by ocean-currents, no Christians have wintered in so high a latitude as this. They are Russian sailors who made the encounter there—men inured to hardships and cold. Our darkness has ninety days to run before we shall get back again even to the contested twilight of to-day. Altogether our winter will have been sunless for one hundred and forty days.

"*December 15th.*—We have lost the last vestige of our mid-day twilight. We cannot see print, and hardly paper; the fingers cannot be counted a foot from the eyes. Noonday and midnight are alike; and, except a vague glimmer in the sky that seems to define the hill outlines to the south, we have nothing to tell us that this Arctic world of ours has a sun. In the darkness, and consequent inaction, it is almost in vain that we seek to create topics of thought, and, by a forced excitement, to ward off the encroachments of disease."

But in due time the long Arctic night passed away, and the season came round for undertaking the sledge journeys which were the main object of the expedition. But Dr. Kane was then met by a new difficulty. Out of the nine splendid Newfoundland and thirty-five Eskimo dogs which he had originally possessed, only six had survived a peculiar malady that had seized them during the winter; and though some fresh purchases were made from the Eskimos who visited Rensselaer Harbour early in April, his means of transport remained wholly inadequate.

Kane, moreover, who though strong of heart was weak of body, had suffered much from the rigour of the climate, and was in a sadly feeble condition when, on the 25th of April 1854, he started on his northward journey. He found the Greenland coast, as he ascended Kane Sea, full of romantic surprises; the cliffs rising to a height of ten hundred and eleven hundred feet, and presenting the boldest and most fantastic outlines. This character is continued as far as the Great Humboldt Glacier. The coast is indented by four great bays, all of them communicating with deep gorges, which are watered by streams from the interior ice-fields. The mean height of the table-land, till it reaches the bed of the Great Glacier, Dr. Kane estimated, in round numbers, at 900 feet; its tallest summit near the water at 1300, and the rise of the background above the general level at 600 more. The face of this stupendous ice-mass, as it defined the coast, was everywhere an abrupt and threatening precipice, only broken by clefts and deep ravines, giving breadth and interest to its wild expression.

THE "THREE BROTHER TURRETS."

Dr. Kane informs us that the most picturesque portion of the coast occurs in the neighbourhood of Dallas Bay. Here the red sandstones contrast very favourably with the blank whiteness, and associate the warm colours of more southern lands with the cold tints of the Arctic scenery. The seasons have acted on the different layers of the cliff so as to give them all the appearance of jointed masonry, and the narrow stratum of greenstone at the top surmounts them with boldly-designed battlements. To one of these "interesting freaks of Nature" Kane gave the name of the "Three Brother Turrets." The crumbled ruin at the foot of the coast-wall led up, like an artificial causeway, to a ravine that blazed at noonday with the glow of the southern sun, when everywhere else the rock lay in blackest shadow. Just at the edge of this lane of light rose the semblance of a castle, flanked with triple towers, completely isolated and defined. These were the Three Turrets.

Still further to the north, a solitary cliff of greenstone, marked by the slaty limestone that once encased it, sprang from a mass of broken sandstone, like the rough-hewn rampart of an ancient city. At its northern extremity, on the brink of a deep ravine, wrought out among the ruin, stood a solitary column, or minaret-tower, the pedestal of which was not less than 280 feet in height, while the shaft was fully 480 feet. Dr. Kane associated this remarkable beacon with the name of the poet Tennyson.

Dr. Kane continued his advance, and on the 4th of May approached the Great Glacier. This progress, however, was dearly earned. Owing to the excessive cold and labour, most of his party suffered from painful prostration; three were attacked with snow-blindness; and all were troubled with dropsical swellings. Off Cape Kent, while taking an observation for latitude, Kane himself was seized with a sudden pain, and fainted. His limbs became rigid. He was strapped upon the sledge, and insisted that the march should be continued. But, on the 5th, he grew delirious, and fainting every time that he was taken from the tent to the sledge, he succumbed entirely.

"My comrades," writes this heroic man, than whom no braver or more resolute spirit ever ventured into the dreary Northern wilds, "would kindly persuade me that, even had I continued sound, we could not have proceeded on our journey. The snows were very heavy, and increasing as we went; some of the drifts perfectly impassable, and the level floes often four feet deep in yielding snow. The scurvy had already broken out among the men, with symptoms like my own; and Morton, our strongest man, was beginning to give way. It is the reverse of comfort to me that they shared my weakness. All that I should remember with pleasurable feeling is, that to five brave men, themselves scarcely able to travel, I owe my preservation."

They carried him back to the brig at Rensselaer Harbour, and for several days he lay fluctuating between life and death. As the summer came on, however, his health slowly improved; and though unable to undertake any sledge excursions in person, he organized a series of expeditions in which his stronger companions took part. Dr. Hayes crossed the strait in a north-easterly direction, reached the opposite coast of Grinnell Land, where the cliffs varied from 1200 to 2000 feet in height, and surveyed it as far as Cape Fraser, in lat. 79° 45′.

He returned on June 1st, and, a few days later, Morton departed to survey the Greenland shore beyond the Humboldt Glacier. His journey was a difficult one, for the obstacles offered by the ice hummocks were sometimes almost insurmountable, and the ice-field was intersected by chasms and water-lanes frequently four feet in width. After skirting the coast of what is now known as Morris Bay, Morton's party came upon easier ground; and presently a long low country opened on the land-ice, a wide plain between large headlands, with rolling hills through it. A flock of brent geese came down this valley, with a whirr of wings, and ducks were seen in crowds upon the open water. Eiders and dove-kies also made their appearance; and tern were very numerous, and exceedingly tame. Flying high overhead, their notes echoing from the rocks, were large white birds, which Morton supposed to be burgomasters. There were also ivory gulls and mollemokes; the former flying very high, and the latter winging their way far out to sea.

The channel (Kennedy Channel) was here unobstructed by ice, and its waves rolled freely and noisily on the shore. Along its verdant margin Morton proceeded warily, and on the 26th of June, 1854, reached the striking headland of Cape Constitution, about 2000 feet in height. Its base was washed by a tremendous surf, through which it was impossible to pass—the *ne plus ultra*, as it seemed, of human enterprise. Climbing from rock to rock, he contrived to reach an elevation of 300 feet; from which he was able to trace the outline of the coast for fifty miles to the north. In the distance rose a range of mountains, very lofty, and rounded at their summits. To the north-west might be seen a bare peak, striated vertically with protruding ridges, and

MORTON ON THE SHORE OF THE SUPPOSED POLAR OCEAN.

soaring to an altitude of between 2500 and 3000 feet. This peak, the most remote northern land then known upon the globe, was named after the great pioneer of Arctic travel, Sir Edward Parry.

The range (Victoria and Albert Mountains) with which it was connected was much higher, Morton thought, than any they had seen on the southern or Greenland side of the bay. The summits were generally rounded, resembling a succession of sugar-loaves and stacked cannon-balls declining slowly in the perspective.

All the sledge-parties were now once more aboard the brig, and the season of Arctic travel had ended. The short summer was rapidly wearing away, and yet the ice remained a rigid and impenetrable barrier. It was evident that the ship could not be liberated, and Kane found himself compelled to decide between two equally dismal alternatives,—the abandonment of the ship, or another winter among the Polar snows. For himself, he resolved to remain; but to those who were willing to venture on the attempt to reach the Danish settlement at Upernavik, he left the choice open. Out of the seventeen survivors of the party, eight, like Dr. Kane, decided to stand by the brig; the others, to push southward to Upernavik. These were provided with all the provisions and appliances that could be spared, and took their departure on Monday, August 28th; carrying with them a written assurance of a brother's welcome should they be driven back—an assurance amply redeemed when severe trials had prepared them to share again the fortunes of their commander.

Dr. Kane confronted the winter with equal sagacity and resolution. He had carefully studied the Eskimos, and concluded that their form of habitation and peculiarities of diet, without their unthrift and filth, were the safest that could be adopted. He turned the brig, therefore, into a kind of *igloë*, or hut. The quarter-deck was well padded with moss and turf, and the cabin below, a space some eighteen feet square, was inclosed and packed from floor to ceiling with inner walls of the same material. The floor itself was carefully calked with plaster of Paris and common paste, and covered two inches deep with Manilla oakum and a canvas carpet. The entrance was from the hold by a low, moss-lined tunnel, the *tossut* of the native huts, with as many doors and curtains to close it up as ingenuity could devise. This was their sitting-room, dining-room, sleeping-room; but there were only ten of them, and the closer the warmer.

DR. KANE PAYING A VISIT TO AN ESKIMO HUT AT ETAH.

While they were engaged in these defences against the enemy, they contrived to open up a friendly intercourse with the Eskimos, visiting them in their snow-huts at the settlements of Etah and Anatoak, distant about thirty and seventy miles from the brig; and, in return for presents of needles, pins, and knives, they undertook to show the white strangers where game was to be procured, as well as to furnish walrus and fresh seal meat. The assistance rendered by the Eskimos was of the greatest value, and we may infer that, without it, Dr. Kane and his followers must have succumbed to the hardships of that dreadful winter.

On the 12th of December, the party which had abandoned the ship suddenly reappeared, finding it impossible to penetrate to the south. They had suffered severely; were covered with rime and snow, and fainting with hunger. It was necessary to use much caution in conveying them below; for after an exposure of such fearful intensity and duration as they had undergone, the warmth of the cabin would have prostrated them completely. They had journeyed three hundred and fifty miles; and their last run from the bay near Etah, some seventy miles in a right line, was through the hummocks with the thermometer at −50°. "One by one," says Kane, "they all came in and were housed. Poor fellows! as they threw open their Eskimo garments by the stove, how they relished the scanty luxuries which we had to offer. The coffee, and the meat-biscuit soup, and the molasses, and the wheat-bread, even the salt pork, which our scurvy forbade the rest of us to touch—how they relished it all! For more than two months they had lived on frozen seal and walrus meat."

We cannot dwell on the various little incidents which marked that sad and terrible winter, but an extract or two from Dr. Kane's journal will show the reader how much the imprisoned explorers endured, and in what spirit they bore their trials:—

"*December 1, Friday.*—I am writing at midnight. I have the watch from eight to two. It is day in the moonlight on deck, the thermometer getting up again to 36° below zero. As I come down to the cabin—for so we still call this little moss-lined igloö of ours—every one is asleep, snoring, gritting his teeth, or talking in his dreams. This is pathognomonic; it tells of Arctic winter, and its companion, scurvy. Tom Hickey, our good-humoured, blundering cabin-boy, decorated with the dignities of cook, is in that little dirty cot on the starboard side; the rest are bedded in rows. Mr. Brooks and myself chock aft. Our bunks are close against the frozen moss-wall, where we can take in the entire family at a glance. The apartment measures twenty feet by eighteen; its height six feet four inches at one place, but diversified elsewhere by beams crossing at different distances from the floor. The avenue by which it is approached is barely to be seen in the moss-wall forward; twenty feet of air-tight space make misty distance, for the puff of outside temperature that came in with me has filled our atmosphere with vesicles of vapour. The avenue—Ben-Djerback is our poetic name for it—closes on the inside with a door well-patched with flannel, from which, stooping upon all fours, you back down a descent of four feet in twelve through a tunnel three feet high, and two feet six inches broad. Arrived at the bottom, you straighten yourself, and a second door admits you into the dark and sorrowing hold, empty of stores, and stripped to its naked ceiling for firewood. From this we grope our way to the main hatch, and mount by a rude stairway of boxes into the open air."

"*February 21, Wednesday.*—To-day the crests of the north-east headland were gilded by true sunshine, and all who were able ascended on deck to greet it. The sun rose above the horizon, though still screened from our eyes by intervening hills. Although the powerful refraction of Polar latitudes heralds his direct appearance by brilliant light, this is as far removed from the glorious tints of day as it is from the mere twilight. Nevertheless, for the past ten days we have been watching the growing warmth of our landscape, as it emerged from buried shadow, through all the stages of distinctness of an India-ink washing, step by step, into the sharp, bold definition of our desolate harbour scene. We have marked every dash of colour which the great Painter in his benevolence vouchsafed to us; and now the empurpled blue, clear, unmistakable, the spreading lake, the flickering yellow; peering at all these, poor wretches! everything seemed superlative lustre and unsurpassable glory. We had so grovelled in darkness, that we oversaw the light.

"Mr. Wilson has caught cold, and relapsed. Mr. Ohlsen, after a suspicious day, startles me by an attack of partial epilepsy; one of those strange, indescribable spells, fits, seizures, whatever name the jargon gives them, which indicate deep disturbance. I conceal his case as far as I can; but it adds to my heavy pack of troubles to anticipate the gloomy scenes of epileptic transport introduced into our one apartment."

"*February 28, Wednesday.*—February closes: thank God for the lapse of its twenty-eight days! Should the thirty-one of the coming March not drag us further downward, we may hope for a successful close to this dreary drama. By the 10th of April we should have seal; and when they come, if we remain to welcome them, we can call ourselves saved.

"But a fair review of our prospects tells me that I must look the lion in the face. The scurvy is steadily gaining on us. I do my best to sustain the more desperate cases; but as fast as I partially build up one, another is stricken down. The disease is perhaps less malignant than it was, but it is more diffused throughout our party. Except William Morton, who is disabled by a frozen heel, not one of our eighteen is exempt. Of the six workers of our party, as I counted them a month ago, two are unable to do out-door work, and the remaining four divide the duties of the ship among them. Hans musters his remaining energies to conduct the hunt. Petersen is his disheartened, moping assistant. The other two, Bonsall and myself, have all the daily offices of household and hospital. We chop five large sacks of ice, cut six fathoms of eight-inch hawser into junks of a foot each (for fuel), serve out the meat when we have it, hack at the molasses, and hew out with crowbar and axe the pork and dried apples, pass up the foul slops and cleansings of our dormitory; and, in a word, cook, *scullionize*, and attend the sick. Added to this, for five nights running I have kept watch from 8 P.M. to 4 A.M., catching cat-naps as I could in the day without changing my clothes, but carefully waking every hour to note thermometers.

"Such is the condition in which February leaves us, with forty-one days more ahead of just the same character in prospect as the twenty-eight which, thank God! are numbered now with the past. It is saddening to think how much those twenty-eight days have impaired our capacities of endurance. If Hans and myself can only hold on, we may work our way through. All rests upon destiny, or the Power which controls it."

It is useless, however, to dwell longer on this melancholy record. Kane saw that to abandon the brig was now the only resource: the ice held it fast, there was no probability of its being released, and a third winter in Rensselaer Bay would have been death to the whole party. As soon, therefore, as the return of spring in some measure recruited the health of his followers, he made the necessary preparations for departure; and on the 20th of May the entire ship's company bade farewell to the *Advance*, and set out on their homeward route. With considerable difficulty and arduous labour they hauled their boats across the rough, hummocky ice, and reached the open sea. On the 17th of June they embarked, and steered for Upernavik, which port they calculated upon reaching in fifty-six days. When they got fairly clear of the land, and in the course of the great ice-drift southward, they found their boats so frail and leaky that they could be kept afloat only by constant bailing; a labour which told heavily on men already weakened with disease and want. Starvation stared them in the face, when happily they fell in with and captured a large seal, which they devoured voraciously; and this opportune help recruited their failing energies. Thenceforth they were in no lack of food, as seals were plentiful; and early in

August, after living for eighty-four days in the open air, they found themselves under the comfortable roofs of Upernavik, enjoying the hospitable welcome of the generous Danes.

Dr. Kane returned to New York on the 11th of October 1855, after an absence of thirty months. His discoveries had been important, his heroism worthy of the race from which he sprung, and none can deny that he had well merited the honours he received. Unfortunately, a frame never very robust had been broken down by the trials of two Arctic winters; and this gallant explorer passed away on the 16th of February 1857, in the thirty-seventh year of his age.

In 1860, Dr. Hayes, the companion of Dr. Kane, took the command of an expedition intended to complete the survey of Kennedy Channel, and to reach, if it were possible, the North Pole. His schooner, the *United States,* was brought up for the winter at Port Foulke, about twenty miles south of Rensselaer Harbour; and early in the following April, Dr. Hayes set out on a sledge and boat journey across the sound, and along the shores of Grinnell Land.

From the eloquent record of his adventures, which does so much credit to his literary skill, "An Arctic Boat Journey," we have already quoted some stirring passages; but the following extract we may be allowed to repeat, on account of the clear light it throws upon the nature of the difficulties Hayes encountered on his northward advance :—

"The track," he says, "was rough, past description. I can compare it to nothing but a promiscuous accumulation of rocks closely packed together, and piled up over a vast plain in great heaps and endless ridges, leaving scarcely a foot of level surface. The interstices between these closely accumulated ice-masses are filled up, to some extent, with drifted snow. The reader will easily imagine the rest. He will see the sledges winding through the tangled wilderness of broken ice-tables, the men and dogs pulling and pushing up their respective loads. He will see them clambering over the very summit of lofty ridges, through which there is no opening, and again descending on the other side—the sledge often plunging over a precipice, sometimes capsizing, and frequently breaking. Again he will see the party, baffled in their attempt to cross or find a pass, breaking a track with shovel and handspike; or, again, unable even with these appliances to accomplish their end, they retreat to seek a better track: and they may be lucky enough to find a sort of gap or gateway, upon the winding and uneven surface of which they will make a mile or so with comparative ease. The snow-drifts are sometimes a help and sometimes a hindrance. Their surface is uniformly hard, but not always firm to the foot. The crust frequently gives way, and in a most tiresome and provoking manner. It will not quite bear the weight, and the foot sinks at the very moment when the other is lifted. But, worse than this, the chasms between the hummocks are frequently bridged over with snow in such a manner as to leave a considerable space at the bottom quite unfilled; and at the very moment when all looks promising, down sinks one man to his middle, another to the neck, another is buried out of sight; the sledge gives way,—and to extricate the whole from this unhappy predicament is probably the labour of hours. It would be difficult to imagine any kind of labour more disheartening, or which would sooner sap the energies of both men and animals."

After encountering difficulties like these, which wore out the strength of most of his party, so that they were compelled to return to the schooner, Dr. Hayes succeeded in crossing the sound, and began his journey along the coast. But the difficulties did not abate, and made such demands on the powers of endurance of the travellers, that the strongest among them broke down, and had to be left behind in charge of another of the party. The resolute Hayes then pushed on, accompanied by Knorr, and on the 18th of May reached the margin of a deep gulf, where further progress was rendered impossible by the rotten ice and broad water-ways. From this point, however, he could see, on the other side of the channel, and immediately opposite to him, the lofty peak of Mount Parry, discovered in 1854 by the gallant Morton; and more to the north, a bold conspicuous headland, which he named Cape Union, the most northern known land upon the globe. Beyond it, he thought he saw the open sea of the Pole, which, from Cape Union, is not distant five hundred miles; but the voyage of the *Polaris*, at a later date, has shown that what he saw was only a land-locked bay.

On the 12th of July, the schooner was set free from the ice, but she proved to be too much damaged to continue her dangerous voyage; and satisfied with having proved that a direct and not impracticable route to the Pole lies up Smith Sound and Kennedy Channel, Dr. Hayes returned to Boston.

It is the opinion, however, of some geographers, though scarcely warranted by ascertained facts, that the Pole may more easily be reached by what is known as the Spitzbergen route. They argue that to the east of this snow-crowned archipelago the influence of the Gulf Stream makes itself felt; and they conclude that this great warm current possibly strikes as far as the Pole itself. It is known that Parry, to the north of Spitzbergen, attained the latitude of 82° 45′; and it is recorded that a Hull whaler, the *True-Love*, in 1837, navigated an open sea in lat. 82° 30′ N., and long. 15° E.; so that she might probably have solved the problem and have gained the Pole, had she continued on her northerly course.

Holding this belief, the illustrious German geographer, Dr. Petermann, succeeded in raising funds for a German expedition in 1868; and the *Germania*, a brig of eighty tons, under the command of Captain Koldewey, sailed from Bergen on the 24th of May, for Shannon Island, in lat. 75° 14′ N., the furthest point on the Greenland coast reached by Sabine in 1823. She was accompanied by the *Hansa*, Captain Hegemann; and both ships were equipped in the most careful manner, and liberally supplied with appliances and stores.

On the 9th of July the expedition was off the island of Jan Mayen, and at midnight on that day was sailing direct to the northward. A heavy fog came on, and the two ships, even when sailing side by side, could not see one another, and communication could be maintained only by the use of the speaking-trumpet. Their crews might then conceive an idea of that impenetrable chaos which, according to Pythias, terminated the world beyond Thule, and which is neither air, nor earth, nor sea. It is impossible to imagine anything more melancholy than this gray, uniform, infinite veil or canopy; ocean itself, far as the eye can reach, is gray and gloomy.

For five successive days the weather remained in this condition, the fog alone varying in intensity, and growing thicker and thicker. On the 14th a calm prevailed, and the *Germania* lowered a boat to pick up drift-wood and hunt the sea-gulls. The ice-blink on the horizon showed that the ships were drawing near the great ice-fields of the Polar Ocean; and another sign of their proximity was the appearance of the ivory gull (*Larus eburneus*), which never wanders far from the ice. Occasionally the ships fell in with a rorqual, or nord-caper, as the seamen call it,—a species of whale distinguished by the presence of a dorsal fin.

On the morning of the 15th of July a light breeze blew up from the south, and the two ships sailed steadily on their north-western course through a sea covered with floating ice. An accustomed ear could already distinguish a distant murmur, which seemed to draw nearer and yet nearer; it was the swell of the sea breaking on the far-off ice-field. Nearer and yet nearer! Everybody gathered upon deck; and suddenly, as if in virtue of some spell, the mists cleared away, and the adventurers saw before them, within a few hundred yards, the ice! It formed a long line, like a cliff-wall of broken and rugged rocks, whose azure-tinted precipices glittered in the sun, and repelled, unmoved, the rush of the foamy waves. The summit was covered with a deep layer of blinding-white snow.

They gazed on the splendid panorama in silence. It was a solemn moment, and in every mind new thoughts and new impressions were awakened, in which both hope and doubt were blended.

The point where the *Germania* had struck the ice was lat. 74° 47′ N. and long. 11° 50′ E., and the icy barrier stretched almost directly from north to south. The *Hansa* touched the ice on the same day, but in lat. 74° 57′ N., and long. 9° 41′ E.

The two ships, which had separated in the fog, effected a union on the 18th, and the *Germania* taking the *Hansa* in tow, they made towards Sabine Island. After awhile, the towing-rope was thrown off, the *Germania* finding it necessary to extinguish her fires and proceed under canvas. They then followed up, in a southerly direction, the great icy barrier, seeking for an opening which might afford them a chance of steering westward.

On the 20th, the *Germania* found the ice so thick in the south-west that she adopted a westerly course, and hoisted a signal for the captain of the *Hansa* to come on board to a conference. The latter, however, misinterpreted it, and instead of reading the signal as "Come within hail," read it as "Long stay a peak;" crowded on all sail, and speedily disappeared in the fog, which grew wonderfully intense before the *Germania* could follow her. Through this curious error the two ships were separated, and for fourteen months the crew of the *Germania* remained in ignorance of the fate of their comrades.

Before following the *Germania* on her voyage of discovery, we propose to see what befell the *Hansa* among the Arctic ice.

Captain Hegemann had understood the signal of his senior officer to mean that the ships were to push on as far as possible to the westward, and, as we have seen, he crowded on all sail. But when the fog closed in, and he found himself out of sight of the *Germania*, he lay-to, in the hope that the latter might rejoin him. Disappointed in this, he kept on his way, and on the 28th of July sighted the rocky and gloomy coast of East Greenland, from Cape Bröer-Ruys to Cape James.

The weather continued fine. By the light of the midnight sun, which illuminated the fantastic outlines of the bergs, the adventurers engaged in a narwhal-hunt. Nothing is more extraordinary than the effect of the rays of the midnight sun penetrating into an ocean covered with floating ice. The warm and cold tones strike against each other in all directions; the sea is orange, leaden-gray, or dark green; the reefs of ice are tinged with a delicate rose-bloom;

THE CREW OF THE "HANSA" TRYING TO LASSO A BEAR.

broad shadows spread over the snow, and the most varied effects of mirage are produced everywhere in the tranquil waters.

On the 9th of September, the *Hansa* found the channel of free water in which she had been

THE MIDNIGHT SUN, GREENLAND.

navigating closed by a huge mass of ice, and to protect her against the drift of the floating bergs she was moored to it with stout hawsers. A few days later, the ice was broken up by a gale of wind from the north-east, and the hawsers snapped. The ice accumulating behind the ship

A BEAR AT ANCHOR.

raised it a foot and a half. On a contiguous sheet of ice, the explorers discovered a she-bear with her cub, and a boat was despatched in pursuit. The couple soon caught sight of it, and began to trot along the edge of the ice beside the boat, the mother grinding her teeth and licking her beard. Her enemies landed, and fired, and the bear fell in the snow, mortally wounded. While the cub was engaged in tenderly licking and caressing her, several attempts were made to capture it with a lasso; but it always contrived to extricate itself, and at last took to flight, crying and moaning bitterly. Though struck with a bullet, it succeeded in effecting its escape.

On the 12th they again saw a couple of bears coming from the east, and returning from the sea towards the land. The mother fell a victim to their guns, but the cub was captured, and chained to an anchor which they had driven into the ice. It appeared exceedingly restless and disturbed, but not the less did it greedily devour a slice of its mother's flesh which the sailors threw to it. A snow wigwam was hastily constructed for its accommodation, and the floor covered with a layer of shavings; but the cub despised these luxuries of civilization, and preferred to encamp on the snow, like a true inhabitant of the Polar Regions. A few days afterwards it disappeared with its chain, which it had contrived to detach from the anchor; and the weight of the iron, in all probability, had dragged the poor beast to the bottom of the water.

SKATING—OFF THE COAST OF GREENLAND.

The *Hansa* was now set fast in the ice, and no hope was entertained of her release until the coming of the spring. Her crew amused themselves with skating, and, when the weather permitted, with all kinds of gymnastic exercises. It became necessary, however, to consider what preparations should be made for encountering the Arctic winter, one of the bitterest enemies with which man is called upon to contend. The *Hansa* was strongly built, but her commander feared she might not be able to endure the more and more frequent pressure of the ice. At first, it was proposed to cover the boats with sail-cloth and convert them into winter-quarters; but it was felt that they would not afford a sufficient protection against the rigour of the Polar climate, its furious winds, its excess of cold, its wild whirlwinds of snow. And therefore it was resolved to erect on the ice-floe a suitable winter-hut, constructed of blocks of coal. Bricks made of this material have the double advantage of absorbing humidity, and reflecting the heat which they receive. Water and snow would serve for mortar; and a roof could be made with the covering which protected the deck of the *Hansa* from the snow.

The ground-plan of the house was designed by Captain Hegemann; it measured twenty feet in length, and fourteen feet in width; the ridge of the roof was eight feet and a half, and the side walls four feet eight inches in elevation. These walls were composed of a double row of bricks nine inches wide up to a height of two feet, after which a single row was used. They were cemented in a peculiarly novel fashion. The joints and fissures were filled up with dry snow, on which water was poured, and in ten minutes it hardened into a compact mass, from which it would have been exceedingly difficult to extract a solitary brick. The roof consisted of sails and mats, covered with a layer of snow. The door was two and a half feet wide, and the floor was paved with slabs of coal. Into this house, which was completed in seven days, provisions for two months were carried, including four hundred pounds of bread, two dozen boxes of preserved meat, a flitch of bacon, some coffee and brandy, besides a supply of firing-wood, and some tons of coal.

On the 8th of October, after the completion of the house, a violent snowstorm broke out, which would assuredly have rendered its construction impossible, and which, in five days, completely buried both the ship and the hut. Such immense piles of snow accumulated on the deck of the *Hansa*, that it was with the greatest difficulty the seamen could reach their berths.

From the 5th to the 14th of October the drift of the current was so strong, that the ice-bound ship was carried no fewer than seventy-two miles towards the south-south-east.

Meantime, the pressure of the ice continued to increase, and the *Hansa* seemed held in the tightening grasp of an invincible giant. Huge masses rose in front, and behind, and on both sides, and underneath, until she was raised seventeen feet higher than her original position. Affairs seemed so critical, that Captain Hegemann hastened to disembark the stores of clothing, the scientific instruments, charts, log-book, and diaries. It was found that through the constant strain on her timbers the ship had begun to leak badly, and on sounding, two feet of water were found in the pumps. All hands to work! But after half an hour's vigorous exertions, the water continued to rise, slowly but surely; and the most careful search failed to indicate the locality of the leak. It was painfully evident that the good ship could not be saved.

"Though much affected," says the chronicler of the expedition, "by this sad catastrophe, we endured it with firmness. Resignation was indispensable. The coal hut, constructed on the shifting ice-floe, was thenceforward our sole refuge in the long nights of an Arctic winter, and was destined, perhaps, to become our tomb.

"But we had not a minute to lose, and we set to work. At nine o'clock P.M. the snow-fall ceased; the sky glittered with stars, the moon illuminated with her radiance the immense wilderness of ice, and the rays of the Aurora Borealis here and there lighted up the firmament with their coloured coruscations. The frost was severe; during the night the thermometer sank to – 20° R. One half the crew continued to work at the pumps; the other was actively engaged in disembarking on the ice the most necessary articles. There could be no thought of sleep, for in our frightful situation the mind was beset by the most conflicting apprehensions. What would become of us at the very outset of a season which threatened to be one of excessive rigour? In vain we endeavoured to imagine some means of saving ourselves. It was not possible to think seriously of an attempt to gain the land. Perhaps we might have succeeded, in the midst of the greatest dangers, in reaching the coast by opening up a way across the ice-floes, but we had no means of transporting thither our provisions; and it appeared, from the reports of Scoresby, that we could not count on finding any Eskimo establishments,—so that our only prospect then would have been to die of hunger."

The sole resource remaining to the explorers was to drift to the south on their moving ice-floe, and confine themselves, meantime, to their coal hut. If their ice-raft proved of sufficient strength, they might hope to reach in the spring the Eskimo settlement in the south of Greenland, or come to gain the coast of Iceland by traversing its cincture of ice.

It was on the 22nd of October, in lat. 70° 50′ N., and long. 21° W., that the *Hansa* sank beneath the ice. Dr. Laube writes: "We made ourselves as snug as possible, and, once our little house was completely embanked with snow, we had not to complain of the cold. We enjoyed perfect health, and occupied the time with long walks and with our books, of which we had many. We made a Christmas-tree of birch-twigs, and embellished it with fragments of wax taper."

To prevent attacks of disease, and to maintain the cheerfulness of the men, the officers of the expedition stimulated them to every kind of active employment, and laid down strict rules for the due division of the day.

At seven in the morning, they were aroused by the watch. They rose, attired themselves in their warm thick woollen clothing, washed in water procured by melting snow, and then took their morning cup of coffee, with a piece of hard bread. Various occupations succeeded: the construction of such useful utensils as proved to be necessary; stitching sailcloth, mending clothes, writing up the day's journal, and reading. When the weather permitted, astronomical observations and calculations were not forgotten. At noon, all hands were summoned to dinner, at which a good rich soup formed the principal dish; and as they had an abundance of preserved vegetables, the bill of fare was frequently changed. In the use of alcoholic liquors the most rigid economy was observed, and it was on Sunday only that each person received a glass of port.

The ice-floe on which their cabin stood was assiduously and carefully explored in all directions. It was about seven miles in circuit, and its average diameter measured nearly two miles.

The out-of-door amusements consisted chiefly of skating, and building up huge images of snow—Egyptian sphynxes and the like.

The borders of the ice-floe, especially to the west and south-west, presented a curious aspect; the attrition and pressure of the floating ice had built up about it high glittering walls, upwards of ten feet in elevation. The snow-crystals flashed and radiated in the sun like myriads of diamonds. The red gleam of morning and evening cast a strange emerald tint on the white surface of the landscape. The nights were magnificent. The glowing firmament, and the snow which reflected its lustre, produced so intense a brightness, that it was possible to read without fatigue the finest handwriting, and to distinguish remote objects. The phenomenon of the Aurora Borealis was of constant occurrence, and on one occasion was so wonderfully luminous that it paled the radiance of the stars, and everything upon the ice-floe cast a shadow, as if it had been the sun shining.

Near the coal-cabin stood two small huts, one of which served for ablutions, the other as a shed. Round this nucleus of the little shipwrecked colony were situated at convenient points the piles of wood for fuel, the boats, and the barrels of patent fuel and pork. To prevent the wind and snow from entering the dwelling-hut, a vestibule was constructed, with a winding entrance.

The greatest cold experienced was – 29° 30′ F., and this was in December. After Christmas the little settlement was visited by several severe storms, and their ice-raft drifted close along the shore, sometimes within eight or nine miles, amidst much ice-crushing,—which so reduced it on all sides, that by the 4th of January 1870 it did not measure more than one-eighth of its original dimensions.

On the 6th of January, when they had descended as far south as 66° 45′ N. lat., the sun reappeared, and was joyfully welcomed.

On the night of the 15th of January, the colony was stricken by a sudden and terrible alarm. The ice yawned asunder, immediately beneath the hut, and its occupants had but just time to take refuge in their boats. Here they lay in a miserable condition, unable to clear out the snow, and sheltered very imperfectly from the driving, furious tempest. But on the 17th the gale moderated, and as soon as the weather permitted they set to work to reconstruct out of the ruins of the old hut a new but much smaller one. It was not large enough to accommodate more than half the colony; and the other half took up their residence in the boats.

February was calm and fine, and the floe still continued to drift southward along the land. The nights were gorgeous with auroral displays. Luminous sheaves expanded themselves on the deep blue firmament like the folds of a fan, or the petals of a flower.

March was very snowy, and mostly dull. On the 4th, the ice-raft passed within twenty-five miles of the glacier Kolberger-Heide. A day or two later, it nearly came into collision with a large grounded iceberg. The portion nearest to the drifting colony formed an immense overhanging mass; its principal body had been wrought by the action of the sun and the waves into the most capricious forms, and seemed an aggregate of rocks and pinnacles, towers and gateways. The castaways could have seized its projecting angles as they floated past. They thought their destruction certain, but the fragments of ice which surrounded the raft served as "buffers," and saved it from a fatal collision.

On the 29th of March, they found themselves in the latitude of Nukarbik, the island where Graab, the explorer, wintered, from September 3rd, 1827, to April 5th, 1830. They had cherished the hope that from this spot they might be able to take to their boats, and start for Friedrichstal, a Moravian missionary station on the south coast of Greenland. However, the ice was as yet too compact for any such venture to be attempted.

For four weeks they were detained in the bay of Nukarbik, only two or three miles from the shore, and yet unable to reach it. Their raft was caught in a kind of eddy, and sometimes tacked to the south, sometimes to the north. The rising tide carried it towards the shore, the ebbing tide floated it out again to sea. During this detention they were visited by small troops of birds, snow linnets and snow buntings. The seamen threw them a small quantity of oats, which they greedily devoured. They were so tame that they allowed themselves to be caught by the hand.

SNOW LINNETS AND BUNTINGS VISITING THE CREW OF THE "HANSA."

From the end of March to the 17th of April, the voyagers continued their dreary vacillation between Skieldunge Island and Cape Moltke; a storm then drove them rapidly to the south. The coast, with its bold littoral mountain-chain, its deep bays, its inlets, its islands, and its romantic headlands, offered a succession of novel and impressive scenes; and specially imposing was the great glacier of Puisortok, a mighty ice-river which skirts the shore for upwards of thirty miles.

Early in May they had reached lat. 61° 12′.

On the 7th, some water-lanes opened for them a way to the shore; and abandoning the ice-raft, they took to their boats, with the intention of progressing southward along the coast. At first they met with considerable difficulty, being frequently compelled to haul up the boats on an ice-floe, and so pass the night, or wait until the wind was favourable. As this necessitated a continual unloading and reloading of the boats, the work was very severe. At one time they

THE CREW OF THE "HANSA" BIVOUACKING ON THE ICE.

were detained for six days on the ice, owing to bad weather, violent gales, and heavy snow-showers. The temperature varied from + 2° during the day to – 5° R. during the night.

Their rations at this period were thus distributed :—In the morning, a cup of coffee, with a piece of dry bread. At noon, for dinner, soup and broth; in the evening, a few mouthfuls of cocoa, of course without milk and sugar.

They were compelled to observe the most rigid economy in the use of their provisions, lest, before reaching any settlement, they should be reduced to the extremities of famine. Yet their appetite was very keen; a circumstance easily explained, for they were necessarily very sparing in their allowance of meat and fat, which in the rigorous Arctic climate are indispensable as nourishment.

As no change took place in the position of the masses of ice which surrounded them, they resolved to drag their boats towards the island of Illiudlek, about three marine miles distant. They began this enterprise on the evening of the 20th, making use of some stout cables which they had manufactured during the winter, and harnessing themselves by means of a brace passed across the shoulders. That evening they accomplished three hundred paces. Snow fell heavily, and melted as fast as it fell, so that during their night-bivouac they suffered much from damp.

The next day they found before them such a labyrinth of blocks and fragments of ice, floating ice-fields, and water-channels, that they were constrained to give up the idea of hauling their boats across it, and resolved to wait for the spring tide—which, they knew, would occur in a few days. The delay was very wearisome. To beguile the time, some of the seamen set to work at wood-carving, while the officers and scientific gentlemen manufactured the pieces for a game of chess. Others prepared some fishing-lines, eighty fathoms long, in the hope of catching a desirable addition to their scanty bill of fare.

On the 24th, the weather was splendid. The sun shone in a cloudless sky, and wherever its genial radiance fell the thermometer marked + 28° 5′ R. This was an excellent opportunity for drying their clothes, which, as well as their linen, had been thoroughly soaked innumerable times. The coverings were removed from the boats, which, in the warm sunshine, exhaled great clouds of vapour. The cook endeavoured to add to his stores of provisions; but the seals churlishly refused to make their appearance, the fish disdained to nibble at the fat-baited hooks, and the stupid guillemots were cunning enough to escape the best directed shots.

M. Hildebrandt, with two seamen, made an attempt—in which they succeeded—to reach the island of Illiudlek, which lay about three miles off, and is from 450 to 500 feet in height. They found it a desert; not a trace of vegetation; its shores very steep, and at some points precipitous; its surface torn with crevasses and ravines. The only accessible part seemed on the north; but as the evening was drawing in, they had no time for exploration, and made haste to return to the boats.

The castaways now came to a resolution to seek a temporary refuge on this desolate isle. As the heat of the sun was sufficient to render their labour very painful, and they suffered much from the effects of the snow upon their eyes, they went to work at night, dragging their boats forward with many a weary effort, and rested during the daytime. In this way they reached the island on the 4th of June.

Here they moored their boats in a small bay sheltered by a wall of rocks from the north wind, which they named *Hansa-Hafen.* Next day they shot two-and-twenty divers, which provided them with a couple of good dinners. The supply was very valuable, as the stock of provisions on hand would not last above a fortnight.

After a brief rest, the adventurers resumed their voyage, keeping close in-shore, and struggling perseveringly amidst ice and stones—and further checked by an inaccurate chart, which led them into a deep fiord, instead of King Christian IV. Sound. On the 13th of June, however, they arrived at the Moravian missionary station of Friedrichstal, where their countrymen received them with a hearty welcome. For two hundred days they had sojourned upon a drifting ice-field, experiencing all the hardships of an Arctic winter, aggravated by an insufficiency of food.

They reached Julianshaab on the 21st of June; embarked on board the Danish brig *Constance;* and were landed at Copenhagen on the 1st of September.

We must now return to the *Germania.*

Captain Koldewey made several bold attempts to penetrate the pack-ice, but proved unsuccessful in all until, on the 1st of August, he reached lat. 74°, where he contrived to effect a passage; and though much delayed by a succession of fogs and calms, he made his way to Sabine Island,—and dropped anchor on its southern side, in lat. 74° 30′ N., and long. 29° W., on the 5th of August.

On the 10th he again passed towards the north, keeping along the Greenland shore until, in lat. 75° 31′ N., his advance was checked by a mass of closely-packed ice, which stretched from the coast of the mainland out to Shannon Island, a long unbroken line of fourteen miles. It presented a very formidable appearance, being edged in some places with a fringe of broken ice, boulders, and blocks, rising in heaps and hummocks forty feet high.

The *Germania* remained in this position for several days. As nothing but ice was visible to the northward, and no prospect opened up of further progress in that direction, Captain Koldewey moved his ship to the south side of the island on the 16th of August, and dropped anchor close to Cape Philip Broke.

Eleven days were spent in a careful exploration of Shannon Island, during which time a musk-ox was shot, and close watch was kept from an elevated point on the ice lying to the northward. But as it continued solid and immovable, and the end of the season was at hand, Captain Koldewey returned northward, and brought his vessel to anchor on the south side of Pendulum Island on the 27th of August.

When it became necessary to make preparations for facing the coming winter, Captain Koldewey moved his ship on the 13th of September into the little harbour he had occupied on the 5th of August. Their subsequent experience showed it to be the only secure one between the parallels of 74° and 77°. A few days later the ship was frozen in.

The first sledging-party was despatched on the 14th of September, and remained out for eight days. After reaching the mainland, they travelled for four days up a newly-discovered fiord, finding many petrifactions and much lignite. They also saw large herds of musk-oxen. Vegetation was abundant, but chiefly composed of species of *Andromeda.* In the course of this excursion, our explorers had one or two adventures with bears. First, a female, with her two cubs, paid them a visit, but being received with some volleys of musketry, quickly beat a retreat. On another occasion, a daring intruder found his way into their tent. His temerity, however, cost him his life; and the Germans banqueted gaily on the fat and flesh with which he incontinently supplied them.

A RASH INTRUDER.

When the winter preparations were completed, Captain Koldewey organized several shooting parties, who made good booty of reindeer and musk-oxen, and added most satisfactorily to the provision-supplies; no fewer than fifteen hundred pounds of good beef and venison attesting the skill and good fortune of the hunters. But after the beginning of November, neither musk-oxen, reindeer, nor bears were visible.

A second sledge journey was undertaken towards the end of October in a southerly direction. The party discovered another fiord, and returned on the 4th of November. On the following day the sun disappeared altogether, and the dreary Arctic night of three months' duration overtook them.

The close of the year was marked by a succession of violent storms, and the temperature rose to 25° F. It soon fell again to zero, however; but it was not until 1870 that it indicated the maximum of cold experienced throughout the winter,—namely, 40° F. Of the December gales, the most furious broke out on the 16th, and lasted until the 20th. It set free the ice in the harbour, and even to within three hundred yards of the ship; but fortunately she had been anchored in the most sheltered part of the bay, and close to the shore, in only ten feet of water; otherwise the crushed-up ice, moving with the currents, would probably have carried her away to almost certain destruction.

The heroic little company, however, were nowise disheartened by the gloom and hardship of their situation. From Captain Koldewey's account, they would seem to have spent a right merry Christmas, after the hearty German fashion. They danced by starlight upon the ice; they cele-

BEAR-HUNTING—GREENLAND.

brated Christmas Eve with open doors, the temperature being 25° F.; with the evergreen *Andromeda* they made a famous Christmas-tree; they decorated the cabin with flags, and spread out upon their tables the gifts prepared for the occasion by kindly hands: each received his share, and each joined in and contributed to the general merriment.

The Yule-tide festivities over, they made ready the equipments for their sledging expeditions in the ensuing spring,—the object of the most important of these being to attain the highest possible degree of north latitude.

In February the sun returned, and with it the bears; and the daily excursions upon the island, undertaken by the scientific members of the expedition, were rendered dangerous by their audacity. Every one was required to go armed, yet some accidents occurred. One of the "scientists" was severely wounded in the head, and dragged upwards of four hundred paces before his comrades rescued him from the bear. After the lapse of a few weeks, however, he recovered from his wounds.

On the 24th of March, the first sledge-party left the ship, and travelled northward until, on the 15th of April, they reached 77° 1′ N. lat. Then the wild northerly gales compelled them to retrace their steps. On their return they were fortunate enough to shoot some bears, whose blubber supplied them with fuel to warm their food; and the wind filling the sails which they had hoisted on their sledges, they progressed with such rapidity as to reach the ship on the 27th of April.

At the northernmost point attained by this party,—lat. 77° 1′,—the belt of land-ice which skirted the shore seemed to the travellers to be four miles in width and several years old. They speak of it as a "bulwark built for eternity." Out to seaward, the ice, which was very hummocky, stretched in an unbroken expanse.

"INTO A WATER-GAP."

Two other sledge-parties were sent out early in May: one of these was employed in making geographical and scientific explorations of the neighbouring coast of Greenland; the other in attempting the measurement of an arc of the meridian. Their journeys were difficult enough and troublesome enough, and made large demands on the energies of those who undertook them. Crossing hummocks and rugged ice was weary work, and sometimes the whole party plunged into deep drifts of snow. On one occasion, the sledge was precipitated into a water-gap, or crevasse; and before it could be recovered and hauled up on the ice-floe, they were compelled to unload it, and remove each article, one by one. Then again they would have to make their way through a storm of pitiless violence; the north wind driving the frozen snow into their faces with a fury that almost blinded them. Up to their knees in the new snow, they pressed forward with a dogged intrepidity; enduring hardships and triumphing over obstacles of which the "mob of gentlemen who stay at home at ease" can form no adequate conception.

The bears now increased in numbers and in boldness, as if they had determined to besiege the small company now left on board the ship. The greatest caution was necessary to prevent accidents; and though several were shot, their death did not appear to terrify the survivors.

The thaw began about the middle of May, and towards the close of the month the sledge-parties were forced to wade through the water which flooded the surface of the sea-ice.

In June, large portions of land-ice were continually breaking off, and much open water could be descried in the south-east. But it was not until the 10th of July that the ice around the ship broke up. Four days later, boating became practicable, and a voyage was made to the Eskimo village on Clavering Island. It ended in disappointment,—the village having been deserted, and the huts having fallen into ruin.

On the 22nd of July, the *Germania* once more steamed to the northward, to renew the attempt of the preceding year. Her boiler-tubes, however, leaked so seriously, that it was evident the boiler would speedily fail altogether. After some delay it was temporarily patched up; and by following a narrow channel between the loose pack-ice and the firm ice-belt of the coast, she contrived to push forward to the north-east cape of Shannon Island, in lat. 75° 29′ N. Here the ice barrier showed itself compact, solid, and insuperable. The *Germania*, therefore, on the 30th of July, made for the southward, and continued her explorations in that direction. The "Mackenzie Inlet," which Captain Clavering discovered in 1823, was found to have disappeared; its place being occupied by a low, flat plain, on which herds of reindeer were pasturing. So unaccustomed were they to the sight of man, and so fearless of danger, that five of them were speedily shot.

On the 6th of August, the *Germania* discovered and entered a broad, deep fiord in lat. 73° 13′. It was perfectly free from ice; but a fleet of huge icebergs was sailing out of it with the current. It was soon noticed that the farther they ascended this picturesque sea-arm, the warmer became the temperature of the air and of the surface water. It threw off several branches, and these wound in and out among lofty mountains. Their declivities were washed by cascades, and their ravines filled with glaciers; so that the prospect thus unexpectedly opened up of the interior of Greenland was singularly romantic and impressive.

Some of the adventurers ascended a mountain 7000 feet in height; but even from this lofty watch-tower no limit could be discerned to the western or principal arm of the fiord. In about 32° W. long. the mountain-range rose, it was ascertained, to an elevation of 14,000 feet. The *Germania* penetrated for seventy-two miles into this remarkable inlet, and reached 26° W. long.; but her boiler acting irregularly, and Captain Koldewey being apprehensive of the consequences if it wholly failed, commenced his homeward voyage on the 17th of August. He re-entered the pack-ice at the mouth of the fiord, and was occupied until the 24th in forcing his way through it,—reaching the open, iceless sea in lat. 72° N. and long. 14° W.

The *Germania*, owing to the uselessness of her boiler, made the rest of her voyage under sail, and arrived at Bremen in safety on the 11th of September, with all well on board. It is worth

THE CREW OF THE "GERMANIA" IN A SNOW-STORM.

notice that, with the exception of two accidental wounds, this interesting expedition was accomplished without any kind of sickness,—a circumstance which speaks highly for the forethought and carefulness of those engaged in equipping and conducting it.

We have been indebted for our brief notice of the voyage of the *Germania* to a paper by Captain Sir Leopold M'Clintock, who sums up its results in a condensed and intelligible form; and to the narratives by Captain Koldewey and his officers, translated by Mr. Mercier, and published under the direction of Mr. H. W. Bates.

The Greenland shore, under the seventy-fifth parallel of latitude, is not the frozen desert which it has hitherto been supposed to be. It is frequented by large herds of reindeer, as many as fifty having been sighted at a time. Musk-oxen were by no means rare, but made their appearance in troops of fifteen or sixteen; while smaller animals, such as ermines and lemmings, were also met with. Birds were not numerous; shoals of walruses were noticed, but no whales.

Geographically speaking, the voyage was valuable from the observations obtained in reference to a region which previously was almost unknown.

The absence of natives, and of all recent traces of them, is a remarkable fact. In 1829, Captain Graah found the northern Greenlanders ranging as high as 64° 15′ N. lat.; but they knew nothing of any human beings living further north; nor could they themselves travel in that direction, the way being blocked up by huge impassable glaciers.

In 1822, when Scoresby partially explored the Greenland coast between the parallels of 70° and 72° 30′, he discovered many ruined habitations and graves, but no recent indications of human beings.

In the following year, Captain Clavering met with a party of Eskimos in 74°; but neither he nor Scoresby found reindeer or musk-oxen; and the fact ascertained by the *Germania* that, in 1869, these animals were numerous, and devoid of any fear of man, gives reason to suppose that few, if any, of this isolated tribe of Eskimos are now in existence. Now, as the musk-oxen, and also the reindeer, seem to have wandered hither from the *northward*, we may conjecture that the natives followed the same route.

"If it be true," says M'Clintock, "that this migration of men and animals was effected from west to east along the northern shore of Greenland, we naturally assume that it does not extend far towards the Pole; that, probably, its most northern point is at the eastern outlet of Kennedy Channel, and that it turns from thence sharply towards the east and north-east,—the distance, in a straight line, to the most northern point reached by Koldewey, is not more than six hundred miles. It is not less strange than sad to find that a peaceable and once numerous tribe, inhabiting a coast-line of at least 7° of latitude in extent, has died out, or has almost died out, whilst at the same time we find, by the diminution of the glaciers and increase of animal life, that the terrible severity of the climate has undergone considerable modification. We feel this saddening interest with greater force when we reflect that the distance of Clavering's village from the coast of Scotland is under one thousand miles! They were our nearest neighbours of the New World."

Returning suddenly to the sixteenth century, we find the names of some Dutch seamen of eminence inscribed in the record of early Arctic Discovery, and amongst these the most illustrious is that of William Barents. We refer to him here, because he is connected with Carlsen's voyage in 1869, which went over much the same ground as that which the Dutch explorer had surveyed nearly three hundred years before.

The merchants of Amsterdam having fitted out a ship—the *Mercurius*, of one hundred tons —to attempt a passage round the northern end of Novaia Zemlaia, the command was given to William Barents; who accordingly sailed from the Texel on the 4th of June 1594.

He sighted Novaia Zemlaia, in lat. 73° 25′ N., on the 4th of July, sailed along its grim, gaunt coast, doubled Cape Nassau on the 10th, and struck the edge of the northern ice on the 13th. For several days he skirted this formidable barrier, vainly seeking for an opening; and in quest of a channel into the further sea, he sailed perseveringly from Cape Nassau to the Orange Islands. He went over no fewer than seventeen hundred miles of ground in his assiduous search,

MATERIALS FOR THE HOUSE.

and put his ship about one-and-eighty times. He discovered also the long line of coast between the two points we have named, laying it down with an exactness which has been acknowledged by later explorers. His men wearying of labour which seemed to yield no positive results, Barents was under the necessity of returning home.

In 1596 the Amsterdammers fitted out another expedition, consisting of two strongly-built ships, under Jacob van Heemskerch and Jan Cornelizoon Rijp, with Barents as pilot, though really in command.

In this voyage the adventurers kept away from the land, in order to avoid the pack-ice, and sailing to the westward, discovered Bear Island on the 9th of June. Then they steered to the northward, and hove in sight of Spitzbergen exactly ten days later. They supposed, however, that it was only a part of Greenland, and were led to bear away to the north-west—a course which was speedily arrested by the eternal icy barrier. Barents then coasted along the western side of

Spitzbergen; and the north-western headland being frequented by an immense number of birds, he called it Vogelsang.

On the 1st of July he again made Bear Island, and here he and Rijp agreed to separate. Of the latter we know only that he was unsuccessful in an attempt to find an opening in the ice on the east of Greenland, and that he returned to Holland in the same year. Of the former the narrative is painfully full and interesting.

Quitting Bear Island, he reached Novaia Zemlaia on the 17th of July, sighting the coast in lat. 74° 40′ N. Keeping along it with characteristic perseverance until the 7th of August, he passed Cape Comfort; but only to find himself once more face to face with the dreary spectacle of the far-reaching Polar ice. It so hemmed and fenced him in on every side, that he was unable to extricate his vessel from it; and being driven into a bay, which he named Ice Haven, "there

ATTACK ON A BEAR.

they were forced, in great cold, poverty, misery, and griefe, to stay all the winter." For the heavy pack-ice drifting into the bay closed it up, and closed around the ship until she was held fast as in iron bonds.

Barents and his sixteen followers now prepared to encounter with a good heart the trials of the long Arctic winter-night. They displayed, in truth, a courage, a patience, and a good fellowship which were heroic. Finding a large supply of drift-wood, they constructed, with the help of planks from the poop and forecastle of the vessel, a sufficiently commodious house, into which they removed all their stores and provisions. They fixed a chimney in the centre of the roof; a Dutch clock was set up and duly struck the weary hours; the sleeping-berths were ranged along the walls; a wine-cask was converted into a bath. All these ingenious devices, however, availed but little against the terrible feeling of depression which is induced by the continuance for so many weeks of a blank and cheerless darkness.

The sun disappeared on the 4th of November, and the cold thereafter increased until it was almost intolerable. Their wine and beer were frozen, and lost all their strength. By means of great fires, by applying heated stones to their feet, and by wrapping themselves up in double fox-skin coats, they barely contrived to keep off the deadly cold. In searching for drift-wood they endured the sharpest pain, and often braved imminent danger. To add to their troubles, they had much ado to defend themselves against the bears, which made frequent assaults on their hut. However, they contrived to slaughter some of the audacious animals, and their fat provided them with oil for their lamps. When the sun disappeared the bears departed, and then the white foxes came in great numbers. They were much more welcome visitors, and being caught in traps, set in the vicinity of the house, supplied the ice-bound voyagers with food and clothing.

When the 19th of December arrived, they found some comfort in the reflection that half

SETTING FOX-TRAPS.

the dreary season of darkness had passed away, and that they could now count every day as bringing them nearer to the joyful spring. They suffered much, but endured their sufferings bravely; and celebrated Twelfth Night with a little sack, two pounds of meat, and some merry games. The gunner drew the prize, and became King of Novaia Zemlain, "which is at least two hundred miles long, and lyeth between two seas."

On the 27th of January every heart rejoiced, for the glowing disc of the sun reappeared above the horizon. But it brought with it their old enemies the bears, against whom they found it necessary to exercise the greatest vigilance.

On the 22nd of February they again saw "much open water in the sea, which in long time they had not seene." During the whole month violent storms broke out, and the snow fell in enormous quantities.

On the 12th of March a gale from the north-east brought back the ice, and the open water disappeared; the ice driving in with much fury and a mighty noise, the pieces crashing against each other, "fearful to hear." Up to the 8th of May the ice was everywhere, and their sad eyes could look forth on no pleasant or hopeful scene. Then it began to break up, and the gaunt, weary explorers prepared to tempt the sea once more. They set to work to repair their two boats, for their ship was so crippled and strained by the ice that she was injured beyond their ability to repair.

On the 14th of June they quitted the place of their long captivity; Barents, before they set out, drawing up in writing a list of their names, with a brief record of their experiences, and depositing it in the wooden hut. He himself was so reduced with sickness, want, and anxiety that he was unable to stand, and had to be carried into the boat. On the 16th, the captain, hailing from the other boat, inquired how the pilot fared. "Quite well, mate," Barents replied;

RELIEVED.

"I still hope to mend before we get to Wardhouse,"—Wardhouse being an island on the coast of Lapland. But he died on the 19th (or, as some authorities say, on the 20th), to the great grief of his comrades, who appreciated his manly character, and placed great reliance on his experience and skill.

The adventurers met with many difficulties from the ice,—sometimes being carried out far from the ice-belt, and at others being compelled to haul the boats for long distances over the rough surface of the floes to reach open water. It has been well observed that there are many instances on record of long ocean-voyages performed in open boats, but that, perhaps, not one is of so extraordinary a character as that which we are describing,—when two small and crazy craft ventured to cross the frozen seas for eleven hundred miles, continually endangered by huge floating ice-masses, threatened by bears, and exposed for forty days to the combined trials of sickness, famine, cold, and fatigue.

At length they arrived at Kola, in Lapland, towards the end of August; and, strangely enough, were taken on board a Dutch vessel commanded by the very Cornelizoon Rijp who had commanded the sister discovery-ship in the previous year. They reached the Maas in safety in October 1597.

No voyager appears to have sailed in the track of Barents, or, at all events, to have visited the place where he wintered, until 1871. No one but he had rounded the north-east point of bleak Novaia Zemlaia. In 1869, however, and on the 16th of May, Captain Carlsen, a Norwegian of much experience in the North Sea trade, sailed from Hammerfest in a sloop of sixty tons, called the *Solid*. On the 7th of September he reached Ice Haven, and on the 9th discovered a rude wooden house standing at the head of the bay. Its dimensions were 32 feet by 20, and it was constructed of planks measuring from 14 to 16 inches in breadth, and 1½ inches thick. These, it was evident, had belonged to a ship, and amongst them were several oak beams. Heaps of bones of seal, bear, reindeer, and walrus, as well as several large puncheons, were collected round the hut. It was the winter-prison of Barents and his companions, and had never been entered by human foot since they had abandoned it. The cooking-pans stood over the fireplace, the old clock hung against the wall; there were the books, and implements, and tools, and weapons which had been of so much service two hundred and seventy-eight years before. It was an Arctic reproduction of the legend of the hundred years' sleep of the fairy princess.

Captain Carlsen gives the following list of articles found in the lone hut on the shore of Novaia Zemlaia:—

Iron frame over the fireplace, with shifting bar; two ship cooking-pans of copper, found standing on the iron frame, with the remains of a copper scoop; copper bands, probably at one time fastened round pails; bar of iron; iron crowbar; one long and two small gun-barrels; two bores or augers, each three feet in length; chisel, padlock, caulking-iron, three gouges, and six files; plate of zinc; earthenware jar; tankard, with zinc lid; lower half of another tankard; six fragments of pepper-pots; tin meat-strainer; pair of boots; sword; fragments of old engravings, with Latin couplets underneath them; three Dutch books; a small piece of metal; nineteen cartridge cases, some still full of powder; iron chest, with lid, and intricate lock-work; fragments of metal handle of same; grindstone; an eight-pound iron weight; small cannon-ball; gun-lock, with hammer and flint; clock, bell of clock, and striker; rasp; small auger; small narrow strips of copper band; two salt and pepper pots, about eight inches high; two pairs of compasses; fragment of iron-handled knife; three spoons; borer; hone; one wooden, and one bronze tap; two wooden stoppers for gun muzzles; two spear or ice-pole heads; four navigation instruments; a flute; lock and key; another lock; sledge-hammer head; clock weight; twenty-six pewter candlesticks and fragments, six in a complete state of preservation; pitcher of Etruscan shape, beautifully engraved; upper half of another pitcher; wooden trencher, coloured red; clock alarum; three scales; four medallions, circular, about eight inches in diameter, three of them mounted in oak frames; a string of buttons; hilt of sword, and a foot of its blade; halberd head; and two carved pieces of wood, one with the haft of a knife in it.

On the 14th of September Captain Carlsen sailed from the Ice Haven, and kept along the east coast of Novaia Zemlaia, encountering bad weather and contrary winds, but succeeding in his chief object, the circumnavigation of the island, which he accomplished on the 6th of October. He returned to Hammerfest early in November.

Our chronological summary now brings us to the Austrian Polar expedition of 1872. The command was intrusted to Lieutenant Payer, an accomplished seaman who had served under Captain Koldewey; Carlsen was engaged as pilot; and the steamer *Tegetthoff* was carefully and abundantly equipped for the voyage. The plan laid down by Lieutenant Payer was well-conceived; namely, to round the north-eastern point of Novaia Zemlaia, and sail eastward until he made the extreme north of Siberia, where he proposed to winter. In the spring, travelling-parties would be sent out on exploring journeys; and the voyage, in summer, would be continued as far as Behring's Strait.

The *Tegethoff* steamed out of Tromsö Harbour on the 13th of July; first fell in with the ice on the 25th, in lat. 74° 15′ N.; and on the 29th sighted the coast of Novaia Zemlaia. Here she was caught in the pack; but steam being got up, repeated charges were made at the enemy, and she was carried bravely into an open water-way, about twenty miles wide, to the north of the Matochkia Strait. On the 12th of August she was joined by the *Isbyörn* yacht, with Count Wilczek and some friends on board. The two vessels anchored close to the shore, in lat. 76° 30′ N., and on the 18th celebrated the Emperor of Austria's birthday. Daily excursions were made by sledge-parties to the adjoining islands, resulting in an accumulation of botanical and geological specimens, besides slaughtered bears and foxes, and quantities of drift-wood. On the 23rd the vessels parted company,—the *Tegethoff* steaming to the northward, and the *Isbyörn* endeavouring to push southward along the coast. On reaching the mouth of the Petchora, Count Wilczek and his friends left her to proceed on the return voyage to Tromsö, while they ascended the Petchora in small boats to Perm, and returned to Vienna by way of Moscow.

The *Tegethoff* spent the winters of 1872 and 1873 in the Icy Sea, and made some discoveries of interest. It returned in safety in the summer of 1874.

We shall close our record with a brief account of the *Polaris* expedition in 1872–73.

Captain Hall, in 1869, returned from a five years' sojourn in the Arctic World, in the course of which he discovered the site of Sir Martin Frobisher's unfortunate colony, collecting numerous highly interesting relics, and gained an intimate acquaintance with the language, habits, customs, and mode of life of the Eskimos. Early in 1870 he conceived the idea of an expedition to reach the North Pole, and obtained from the American Government a wooden river gun-boat of 387 tons, which was re-named the *Polaris*, and from Congress a grant of 50,000 dollars. As Captain Hall was not a seaman, the naval charge of the expedition was placed in the hands of Captain Buddington, a veteran whaler. Dr. Bessels superintended the scientific department; Mr. Meyer went out as meteorologist; and the *personnel* also included Morton, Dr. Kane's steward; Hans the Eskimo, who had served in the expeditions of Kane and Hayes; and Joe and Hannah, with their daughter Silvia, who had accompanied Hall on his return to the United States.

On the 26th of June 1871, Hall was received by the American Geographical Society at New York, where he announced his intention of pushing up Jones Sound, unless he was impeded by heavy pack-ice,—in which case he would attempt the Smith Sound route, only he proposed to take the western coast. Dr. Kane, it will be remembered, took the eastern side. Mr. Grinnell, who laboured so zealously in the search after Franklin, presented Captain Hall with the flag which Admiral Wilkes had carried on his Antarctic expedition, and which had since accompanied De Haven, Kane, and Hayes in their North Polar adventures. "Now I give it to you, sir," said Mr. Grinnell; "take it to the North Pole, and bring it back in a year from next October."

The *Polaris* went on her way, and quitted the northernmost Danish settlement on the Greenland coast in August 1871. Renouncing his intention of penetrating into Jones Sound, Captain Hall entered Smith Sound, and took the *Polaris* a distance of two hundred and fifty miles up the strait leading to the North Pole,—reaching a higher latitude than ships had ever before attained, and ascending within thirty miles of the furthest point ever reached by civilized man. In other words, he sailed through Smith Sound, across Kane Sea, through Kennedy Channel (which he himself discovered), and up a strait which he named after Mr. Robeson, the secretary to the U.S. Navy; finally attaining a latitude of 82° 16′ N. on the 30th of August. Here he was checked by the ice; but he could see a water-horizon to the north-east. The lofty eastern shore, at the furthest visible point, seemed to bend off to the north-east; the western land continued to strike in a northerly direction.

Captain Hall resolved to winter in a small harbour, which he called "Thank God" Bay, in lat. 81° 38′ N., and long. 61° 44′ W. A large inlet, twenty miles wide, and of unascertained depth, which he named the "Southern Fiord," breaks the western coast-line of Polaris Bay. On the 10th of October, Captain Hall, with Mr. Chester, the first mate, and the Eskimos Joe and Hans, set out on a sledge-journey; but could not get beyond the 82nd parallel, to a point in Robeson Strait which he named Newman Bay. On his return, the gallant explorer was seized with illness induced by his arduous exertions, became partially paralyzed, and died on the 8th of November. He was buried on the shore of the strait discovered by his chivalrous enterprise; and there he lies in the heart of the Arctic solitude, another martyr in the cause of Science.

The *Polaris* crew, wintering in lat. 81° 38′ N., found the climate much more genial than it is several degrees further south. In June, the low ground bordering on Thank God Bay was free from snow; rabbits and lemmings abounded; and numerous herds of musk-oxen pastured on the creeping herbage yielded by a friendly soil. The scene was enriched with a glorious burst of wild flowers, and the silence was pleasantly broken by a constant whirr of wings coming from the south. Traces of Eskimos were met with. It was observed that a current of a knot an hour flows down Robeson Strait from the north, and sweeps the ice through Smith Sound, out into Baffin Bay—the great basin into which the North pours its superfluous frozen masses, its huge bergs, and its shattered and shivered ice-floes.

On the death of Hall, the command-in-chief devolved upon Captain Buddington, who seems to have had no appetite for geographical discovery, but instantly to have determined on a homeward voyage.

The *Polaris* left her winter quarters on the 12th of August 1872, and sailed for the south. On reaching lat. 80° 2′ N., she was again imprisoned in the ice; but the current drifted her out into Baffin Bay. On the 15th of October, in lat. 77° 53′ N., she was a second time beset; and the "nip" was so severe that boats and provisions were got upon the ice, and the necessary preparations made to abandon her. However, the floe eased off and relaxed its pressure, and the *Polaris* righted. Unfortunately, Tysen (the second master), Meyer (the meteorologist), the steward, cook, six seamen, and eight Eskimos (men, women, and children), or about half the ship's company, were left upon the ice, with the boats and provisions; and it was impossible for the *Polaris* to rescue them. However, the ice-floe slowly drifted the castaways into safety; and they sheltered themselves from the weather in a snow-cabin, while their rations were augmented by large supplies of birds and seals. Latterly, when the floes began to break up, their supplies grew precarious, and the little company suffered severely from hunger. On April 21st they made good cheer, however, for the Eskimos shot a bear; and on the 29th they were picked up by the *Tigress* steamer, in lat. 53° 35′ N., and only forty miles from land, near Wolf Island.

Meantime, the *Polaris*, driven northward by a violent gale, became unmanageable, and ran ashore at Lyttleton Island, near the mouth of Smith Sound. Here her fourteen seamen and officers passed a second winter in the Arctic World. They did not undergo much privation, happily, for their stock of provisions was abundant, and they received constant assistance from the friendly Eskimos.

In June 1873 the little company built a couple of boats, and proceeded to the south—until, in Melville Bay, they were picked up by the *Ravenscraig* whaler. Thence they were afterwards transferred to the *Arctic* whaler, which carried them safely to Dundee. The United States steamer *Jumata* had already, from information supplied by the other company of explorers, been despatched to Disco in search of the *Polaris*; and the *Tigress*, a vessel specially built for ice-navigation, was sent on the same errand. But, as we have seen, their services were not needed; and both expeditions returned to America in the same season.

The voyage of the *Polaris* enables us to draw some important conclusions. "We now know," says Mr. Markham, "that the American vessel commanded by Captain Hall passed up the strait, in one working season, for a direct distance of two hundred and fifty miles, without a check of any kind, reaching lat. 82° 16′ N.; and that at her furthest point the sea was still navigable, with a water-sky to the northward." The *Polaris*, however, was nothing better than a river-steamer of small power, ill adapted for encountering the perils of Arctic navigation,—with a crew, all told, of thirty men, women, and children, including eight Eskimos. If she could accomplish such a voyage without difficulty, and could attain so high a latitude, it may reasonably be anticipated that a properly equipped English expedition, under equally favourable circumstances, will do, not only as much, but much more, and carry the British flag into the waters of the circumpolar sea. With this view, the Admiralty have fitted out the *Alert* and the *Discovery*, under Captains Nares and Markham. Let us hope that it is reserved for these gallant officers and their companions to solve the great problem of the Pole, and crown the geographical enterprise of Britain with this last and most glorious triumph.

INDEX.

Agaricus muscarius, 139.
Agassiz, experiments of, 116.
Alectoria jubata, 137.
Aletsch glacier, the, described, 47.
Algol, or Medusa's Head, 30.
Arctic Highlanders, 12.
Arctic night, the, characteristics of, 32–34, 93–95.
Arctic region, extent of, 13, 14.
Atmospheric phenomena, 31.
Auk, the, described, 97, 98.
Aurora Borealis, the, phenomena of, 27; theory of, 29.

Baffin, discoveries of, 227.
Barents, adventures of, 266–269.
Barrens, the, region of, described, 16.
Bear, Polar, natural history of the, 85–93; hunting seals, 86; voracity of, 88; affection for its young, 88, 89.
Beechey, Captain, quoted, 45, 46, 55.
Bell, quoted, 78.
Bellot, Lieutenant, quoted, 123.
Bennet, Stephen, voyage of, 228.
Berkeley, quoted, 140.
Birds, migrations of, 11.
Boötes, constellation of, 40.
Bremer, Frederika, quoted, 138.
Brewster's, Sir David, experiment with polarised light, 111.
Burrough, Stephen, voyage of, 222, 223.
Button, Captain, voyage of, 227.
Bylot, Robert, voyage of, 227.

Carlsen, Captain, voyage of, 270.
Cladonia rangiferina, 137.
Clarke, Dr., quoted, 203.
Clavering, Captain, referred to, 12.
Cochlearia, or scurvy-grass, uses of, 139.
Coleridge, quoted, 60.
Constellations, northern, list of, 37.
Cooley, Mr., quoted, 230.
Corvidæ, the, natural history of, 160.
Cryptogamous plants of the north, 141, 142.
Crystallization, process of, 108.
Cygnus musicus, 161.

D'Almeida, Count, quoted, 205, 206.
Davis, Captain John, voyages of, 224, 225.
Dolphins, the, natural history of, 82, 83.

Dorothea, the, narrow escape of, 55.
Dufferin, Lord, quoted, 166, 167, 172.

Eider ducks, the, natural history of, 103; in Iceland, 104.
Eskimo dog, the, description of, 190.
Eskimo, the, hunting the walrus, 68, 69; hut of, 76.
Eskimo kayack described, 182.
Eskimo seal-hunt, an, described, 77, 78.
Eskimo sledge, the, described, 192–196.
Eskimos, the, boundaries of, 175; character, manners, customs, and clothing, 179–196.

Falcon, the, natural history of, 160.
Faraday, ingenious experiment of, 111.
Felinska, Madame, quoted, 211.
Fish in the Arctic seas, 106.
Flora of the Arctic lands, 19.
Fox, Luke, voyage of, 228.
Fox, the Arctic, natural history of, 151–153.
Fox-trap, a, described, 152.
Franklin, Sir J., overland journey of, 231; last expedition of, 231; relics of, discovered, 233, 235, 236.
Fritallaria sarrana, the, properties of, 142, 143.
Frobisher, Sir Martin, voyage of, 223.
Frobisher Strait discovered, 223.

Gale, an Arctic, described, 70.
Gårds of Lapland, described, 207.
Germania, the, expedition of, 245–265.
Geysers, the, phenomena of, 165–167.
Gilbert, Sir Humphrey, death of, 224.
Glacier-ice, peculiarities of, 112.
Glacier in Smith Strait, 124, 127; of Sermiatsialik, 128, 129.
Glaciers, characteristics of, 47; motion of, 113–115; phenomena connected with, 115–118; of the Arctic regions, 118–133; of Spitzbergen, 120, 123.
Godhav'n, 167.
Grampus, the, natural history of, 83, 84.
Greenland, scenery on the coast of, 22.
Guillemot, the, described, 96, 97.
Gull, the, described, 103.

Hall, Captain, expedition and death of, 271, 272.
Hansa, the, voyage and loss of, 245–251.

Hans the Hunter, 186–189.
Hare, the Arctic, 154.
Hartwig, Dr., quoted, 15.
Hayes, Dr., quoted, 25, 35, 43, 49, 50, 59, 87, 95, 96, 124, 125, 127, 128, 138, 152, 153, 186, 191, 192, 244; Arctic expedition of, 244, 245.
Hearne, quoted, 183.
Hecla, the, and the *Fury*, danger of, 56.
Hegemann, Captain, 245.
Hekla, eruption of, 164.
Henderson, Dr., quoted, 172.
Hill, Mr., quoted, 213, 214.
Hobson, Lieutenant, discovers Franklin relics, 234–236.
Holland, Mr., quoted, 173.
Hore, Mr., voyage of, 222.
Hudson, Henry, discoveries of, 225–228.
Humboldt Glacier, description of, 131–134, 238.
Hutchinson, Captain, quoted, 205.

Icebergs, their dimensions, 41; their magnificent appearance, 42, 43; danger to navigation from, 44, 123; breaking up of, 49; range of, 50, 51; in Baffin Bay, 124.
Ice-fields, extent and character of, 54, 56, 57.
Ice-flowers, characteristics of, 108.
Iceland, dimensions of, 162; history of, 162; volcanoes of, 163; dreary landscapes of, 164; geysers of, 166, 167; houses and churches of, 170; travelling in, 172; horses of, 173.
Iceland moss, uses of, 138, 139.

Jacobshav'n, 168.
Jakut merchants, the, enterprise of, 217.
Jakuts, the, as bear-hunters, 212; manners and customs of, 216, 217.
James, Captain Thomas, voyage of, 228.

Kamtschatka, fisheries of, 212; the dog of, 214, 215.
Kamtschatkans, the, characteristics of, 213, 214.
Kane, Dr., quoted, 15, 32, 33, 34, 42, 68, 73, 74, 88, 91, 92, 131–134, 149, 153, 184, 185, 237, 239, 242, 243; Arctic explorations of, 237–244.
Knots, the, habits of, 11, 12.
Koldewey, Captain, referred to, 12; voyage of, 245.

Lagopus, the, 161.
Lamont, Mr., quoted, 60, 62.
Lapland, divisions and extent of, 197; climate of, 197; inhabitants of, 197; the reindeer in, 200; sledging in, 201; an interior in, 204, 205.
Lapp dialect, the, 206.
Lapp hunters, the, boldness of, 202.
Lapps, the, dress, manners, and customs of, 198.
Lapps, the Mountain, character of, 199, 200.
Lapps of West Bothnia, 206, 207.
Laube, Dr., quoted, 252.
Lemming, the Arctic, 154.

Macmillan, Dr., quoted, 135, 136, 137, 141.
Markham, C., quoted, 10, 175, 225, 226, 273.
Marten, the, 155.
Martins, M. Charles, quoted, 119, 120, 121.
M'Clintock, Captain Sir Roderick, quoted, 148, 151, 181, 235; voyage of, 234.
M'Clure, Sir Robert, quoted, 81; discovers North-West Passage, 233.
Mecham, Captain, quoted, 146, 150.
Mer de Glace of Greenland, 127, 128.
Merganser, the, natural history of, 99.
Milton, Lord, and Dr. Cheadle, quoted, 158-160.
Moonlight in the Polar World, 26.
Moraines, described, 115.
Moravian mission-stations in Greenland, 179.
Mosses in the Arctic regions, 139.
Musk-ox, the, natural history of, 149, 150.
Mustelidæ family, the, in the Arctic regions, 155.

Narwhal, the, natural history of, 82.
Newfoundland colonized, 224.
North-West Passage, utility of, 9.
Novaia Zemlaia, temperature of, 21.

Osborn, Admiral Sherard, quoted, 44, 81, 84, 85, 87, 146, 151, 232, 236, 237.
Ostiaks, the, manners and customs of, 211, 212.
Ostrownoje, trade at, 220.
Oxyria, the, uses of, 141.

Pack-ice, description of, 53.
Parry, Captain, quoted, 44, 46, 56, 230; voyages of, 228, 229, 230.
Payer, Lieutenant, voyage of, 270, 271.
Penny, Captain, voyage of, 232-234.
Phænogamous plants of the north, 141.
Phocidæ, the.—See Seal.
Plant-life of Spitzbergen, 142; of Kamtschatka, 142, 143.
Pleiads, the, 39.
Polaris, the, voyage of, 271-273.
Polecat, the, in the Arctic regions, 156, 157.
Pole-Star, the, position of, 36.
Poole, Jonas, voyage of, 228.
Puffin, the, natural history of, 99.

Quenes, or Finns, the, 206.

Rae, Dr., finds relics of Franklin, 233.
Red snow, phenomenon of, explained, 135.
Refraction, phenomena of, 31.
Regelation, what it is, 111.
Reikiavik, description of, 168, 169.
Reindeer, the, natural history of, 144; usefulness of, 145; food of, 146; and wolves, 147; in Siberia, 218, 219.
Reindeer moss, 137.
Rendu, Bishop, quoted, 114.
Richardson, Sir J., quoted, 145.
Rock-hair, 137.
Rorqual, the, 80.
Ross, Sir James, quoted, 145.
Ross, Sir John, voyages of, 228, 231.

Sabine, Sir Edward, quoted, 10.
Sable, the, natural history of, 156.
Samojedes, the, superstitions of, 208, 209; manners and customs of, 210, 211.
Schaitan, an Ostiak idol, 211.
Scoresby, Dr., quoted, 44, 106, 186.
Seal, the, natural history of, 71-73; flesh of, 73, 74; different genera of, 75.
Sermitsialik, glacier of, 127, 128.
Shepherd, Mr., quoted, 104.
Skaptá Jokul, eruption of, 165.
Smew, the, natural history of, 100.
Smith Sound, route by, 228.
Snow, formation of, 108.
Snow-crystals, described, 109.
Snow-line, limit of, 20.
Somerville, Mrs., quoted, 30, 107.
Southey, quoted, 136.
Sporidesmium lepraria, 140.
Spring in the Arctic regions, 34.
Starakis, the, described, 98.
Summer in the Arctic regions, 36.
Swan, the wild, natural history of, 105; the whistling, 161.

Tadebtsios, or Samojede demons, 209.
Tadibe, the Samojede priest, 209.
Tchuktche, the, manners and customs of, 220.
Temperature of Arctic winter, 33.
Tennyson, quoted, 105.
Thingvalla, the, in Iceland, 168.
Thorne, Dr. Robert, Arctic exploration proposed by, 222.
Tripe de roche, 137, 138.
Tundras, the stony, described, 15, 16.
Tungusi, the, characteristics of, 219, 220.
Tyndall, Professor, quoted, 47, 48, 108, 109-111, 112, 113, 115, 117, 118.

Unknown Region, extent of, 10.
Upernavik, described, 176.
Ursa Major, constellation of, 36; description of, 37, 38.

Waigatz, island of, 208.
Walrus, the, natural history of, 63; courage of, 64; gradual decay of, 67.
Walrus-hunt, a, described, 68, 69.
Walrus-hunting, how carried on, 60; proceeds of, 62.
Ware, quoted, 37, 38.
Whale, the, natural history of, 78; characteristics of the Greenland, 79, 80; the Razor-backed, 80.
Whalebone, what it is, described, 79.
Whale-fishery of the Eskimos, 81.
Whirlwinds of the north, 31.
Willoughby, Sir Hugh, loss of, 222.
Wolf, the Arctic, natural history of, 148.
Wolverine, the, cunning of, 157; anecdotes of, 158-160.
Wooded zone of the Arctic regions, 143.
Wrangel, Admiral von, quoted, 20, 81, 218, 221.

Yakutsk, temperature of, 20.

www.ingramcontent.com/pod-product-compliance
Lightning Source LLC
Chambersburg PA
CBHW021515210326
41599CB00012B/1265

* 9 7 8 3 3 3 7 2 2 8 7 9 8 *